NATIONAL
GEOGRAPHIC

FIELD GUIDE TO THE
Water's
Edge

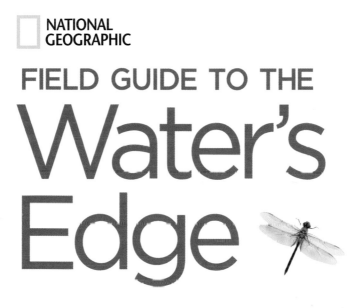

NATIONAL GEOGRAPHIC

FIELD GUIDE TO THE
Water's
Edge

STEPHEN LEATHERMAN
and JACK WILLIAMS

NATIONAL GEOGRAPHIC
WASHINGTON, D.C.

Contents

Pages 2-3: Palm trees sway at sunset on Secret Beach in south Maui, Hawaii.
Opposite: An azure blue sky and turquoise ocean envelop a small island in the Florida Keys.

ABOUT THIS BOOK

Welcome to National Geographic's Field Guide to the Water's Edge, *a book designed for those who enjoy the abundant natural wonders to be discovered along North America's ocean beaches, lakeshores, riverbanks, streams, ponds, and marshes. Here you will find the most common and most fascinating creatures of both saltwater and freshwater landscapes, from sharks and dolphins of the ocean to toads and salamanders of the swamps and springs.*

Your guides on this adventure are two veteran scientists and water's edge experts: Stephen Leatherman, better known as Dr. Beach, and Jack Williams, longtime weather editor of *USA Today.*

Stephen Leatherman, a professor and director of the Laboratory for Coastal Research at Florida International University in Miami, has written or edited 15 books. Widely recognized as an expert on how coastal storms affect beaches, he is probably better known for his annual announcement of America's Top Ten Beaches, made every Memorial Day and now going into its 22nd year.

Jack Williams, the founding weather editor of *USA Today's* newspaper and website, has written six books, including *The Weather Book,* published by *USA Today,* and *The AMS Weather Book: The Ultimate Guide to America's Weather,* published by the American Meteorological Society and the University of Chicago Press. He is an active member of the Explorers Club in Washington, D.C.

This is the second book co-authored by Leatherman and Williams. They also wrote *Hurricanes: Causes, Effects, and the Future,* published in 2008 by Voyageur Press.

Stephen Leatherman   Jack Williams

This book is divided into three parts: ■ **1)** An **introduction** to the fascinating worlds to explore at the water's edge, ■ **2)** A **field guide** to the animals, plants, and other curiosities to be found at the water's edge ■ **3)** A **list of great places** throughout North America, recommended by Dr. Beach and guaranteed to offer enjoyable water's edge experiences

The field guide is the core of the book. Its contents are organized by biological groupings, not by habitat. Sometimes freshwater and saltwater species sit side by side on the same page, indicated by an ▣ for freshwater and an **S** for saltwater. Species that live in both habitats or in brackish regions, where fresh water and salt water mix, are given both letters: **F S**

Occasionally, sensory clues that help us find or identify species will appear in "Look for" and "Listen for" boxes. Circumstances that may be dangerous—plants that cause rashes or animals that sting, for example—warrant "Safety Tip" boxes. And extra tidbits of information are set in "Did you know?" boxes.

Each chapter also contains two "On Location" features that highlight

places selected by our authors for their special landscapes, flora, and fauna. Locator maps show nearby surroundings, and all 16 destinations are located on the map of North America on pages 250–251.

The North America map also locates the 35 places recommended by Dr. Beach on his list of the best water's edge destinations of all time: All-time Top 20 Beaches, Top 10 River Sites, and Top 5 Great Lakes Beaches. Each destination gets a short description in the third section of this book.

FRESHWATER OR SALTWATER?

SENSORY CLUES

EXTRA INFORMATION

THE WATER'S EDGE

On the Edge

each vacations, riverside campouts, lake-
B *shore rambles. For many, recreational time spent in nature involves a visit to the water's edge. Why is it that we are drawn so hypnotically to bodies of water? What do we find when we visit the water's edge? This book may not answer the first of these questions, but it will do much to answer the second. It offers a guide to the natural wonders at the water's edge, whether at the ocean, alongside a lake or river, or near a pond or stream.*

WHERE LAND AND WATER MEET

Something deep within pulls us toward the water. Earth's first life, more than 3.5 billion years ago, lived and evolved in water until around 475 million years ago, when plants and then animals slowly began inhabiting land and evolving into the myriad forms of life that surround us today.

Life Along the Edge
Every known form of life needs water. Plants depend on water in the process of photosynthesis, by which they manufacture the sugars necessary for growth and maturation. Animals not only drink water but also eat water-dependent plants and other animals. The water's edge teems with life, and you are likely to see many kinds of flora and fauna interacting at ocean beaches, riverbanks, lakeshores, ponds, streams, swamps, bogs, and marshes.

Some plants and animals dwell in the water; others begin their lives in the water and mature on land. In addition to the many marine animals and amphibians

A young boy runs along the packed sand beach at Pistol River, Oregon.

A canoeist in Alaska's Denali National Park

Tubing along a rocky woodland stream

that split their lives between land and water, you can expect to see land animals that come to the water's edge to drink and to feed.

In the search for food and water, early humans were drawn to waters' edges. Most ancient cities sprang up near major bodies of water, and the oldest neighborhoods of today's cities still cluster on the waterfront. Even today, when transportation and other infrastructures bring food and water inland, populations along

Ancient Cities on the Water's Edge

Name history's greatest cities. Jericho, Babylon, Alexandria, Athens, Rome, Paris, London, Rio de Janeiro. Every one of them grew up along the water's edge. Luoyang, considered China's oldest city, sits at the confluence of the Luo and Yi Rivers. Through Varanasi, India's oldest city, the sacred Ganges flows.

Now, name the great cities of North America. Montreal, Quebec, Boston, New York, Vancouver, Portland, San Francisco, Los Angeles. In the middle of the continent, you have Chicago, St. Louis, New Orleans. Before the automobile, people and goods moved fastest along the waterways, and trade happened from port to port, river city to river city. Both nature and culture throng along the water's edge.

the coastlines are growing. Between 1980 and 2003, according to the National Oceanic and Atmospheric Administration (NOAA), the populations of coastal counties in the United States grew by 28 percent.

Drawn to the Water

Millions of people throughout North America travel to the water's edge for fun, sports, and hobbies—swimming, snorkeling, scuba diving, surfing, windsurfing, fishing, motorboating, sailing, rowing, canoeing, kayaking, hunting, bird-watching, nature photography, shell collecting . . . The list goes on and on.

We visit bodies of water for many reasons, but, at the core of all these activities, we find nature in all its serenity, power, diversity, and life-forms. National Geographic's *Field Guide to the Water's Edge* is a companion to those adventures.

A family enjoys a day of fun at one of Florida's many lakes.

Kayakers paddle through a natural ice bridge on Bear Lake in Kenai Fjords National Park, Alaska.

WATER, WATER EVERYWHERE

If water were not so familiar, it would amaze us. It is the only natural substance that exists as a gas, a liquid, and a solid at the temperatures and atmospheric pressures found on Earth. As water molecules move at different speeds, the substance moves through different states:

■**Condensation** Vapor molecules go slowly enough that they become liquid.

■**Evaporation** Liquid molecules go fast enough that they escape as vapor.

■**Freezing** Liquid molecules go slowly enough that they latch on to ice as crystals.

■**Melting** Ice molecules vibrate fast enough to break away from the ice and join the liquid.

■**Deposition** Vapor molecules go slowly enough to stay with ice as crystals.

■**Sublimation** Ice crystals vibrate fast and fly off as vapor—water, water everywhere.

Heating a solid, liquid, or gas increases the speeds of molecules. Thus, warming up air or water encourages faster evaporation, while cooling water down speeds up freezing, and cooling the air speeds up condensation and deposition.

Why does all this matter? Because all known forms of life need water, and water from oceans circulates by means of evaporation, becoming the water vapor that winds carry inland, where it condenses and falls as rain or snow. Without evaporation, Earth would have no underground water to supply springs and wells.

THE GLOBAL WATER CYCLE

All of the world's water is part of a global cycle that includes vapor, liquid, and ice. Vapor from bodies of water can rise high into the atmosphere and travel hundreds of miles before falling as rain or snow. That water soaks into the ground and can surface in streams, rivers, and lakes, from which it flows back into the ocean.

As they respire, plants take up water, some from the atmosphere and some from underground. Groundwater, stored in spaces between rocks and soil particles, may seep into rivers or lakes, emerge in a spring, or be pumped from a well for human use.

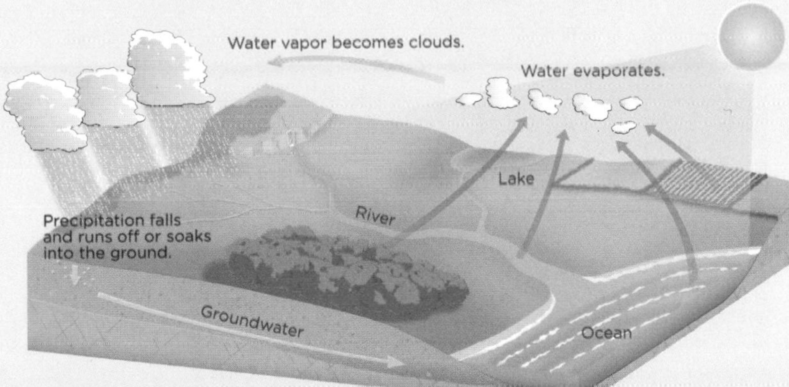

Water vapor becomes clouds.

Water evaporates.

Lake

Precipitation falls and runs off or soaks into the ground.

River

Groundwater

Ocean

SALT, FRESH, OR IN BETWEEN?

The ocean is salty; most lakes (except the Great Salt Lake) and rivers are not. You can taste and sometimes even smell the difference. Ocean water contains many materials, including salts—in fact, you could say that the water in the ocean is a solution of everything on Earth. As water trickles down and through rock, soil, and humus, it picks up particles and chemicals. Volcanic eruptions spew minerals from inside Earth, and they fall into ocean water. Materials from the ocean floor and from marine plants and animals add to the mix.

Of the constituents in ocean water, roughly 85 percent are sodium and chloride, the components of table salt. In contrast, only 16 percent of the materials carried in lake and river water are sodium and chloride. This comparison is one way of explaining the difference between salt water and fresh water. In fact, it's not that one has salt and the other does not; it is a matter of degree. Evaporating ocean water leaves its salt behind.

A waterfall tumbles near California's Big Sur coast.

Sailboats catch the wind off a cape.

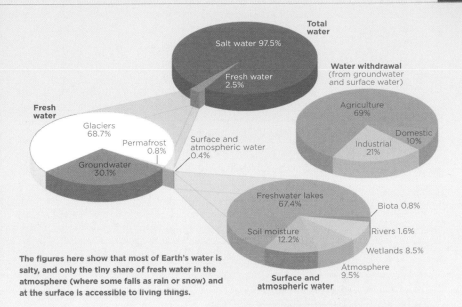

Total water

Salt water 97.5%

Fresh water 2.5%

Water withdrawal
(from groundwater and surface water)

Agriculture 69%

Domestic 10%

Industrial 21%

Fresh water

Glaciers 68.7%

Permafrost 0.8%

Surface and atmospheric water 0.4%

Groundwater 30.1%

Freshwater lakes 67.4%

Biota 0.8%

Soil moisture 12.2%

Rivers 1.6%

Wetlands 8.5%

Atmosphere 9.5%

Surface and atmospheric water

The figures here show that most of Earth's water is salty, and only the tiny share of fresh water in the atmosphere (where some falls as rain or snow) and at the surface is accessible to living things.

Likewise, the boundary between Earth's fresh water and salt water is blurry, because there are regions where changing tides and currents push saltier water up into riverbeds or pull less salty water down into bays and oceans. Called estuaries, these areas are special water's edge locales where saltwater life-forms and freshwater life-forms coexist.

Water Flows Through the Ages

Nature neither creates nor destroys water. The water you drink today is the same water that the dinosaurs drank millions of years ago. Earth and its atmosphere recycle water over and over, and materials dissolved in the water—whether the salt in the ocean or minerals that water picks up when seeping through the soil—are carried and then left behind in a natural process.

Disruptions to the global water cycle can affect everything that lives in the water, along its edges, and even miles away from it. Droughts and floods are nature's prime natural disruptions. Human activities, such as building dams or pumping up groundwater faster than the natural cycle can replace it, can be disruptions too.

WHAT IS A BEACH?

The beach is the quintessential water's edge. The technical definition of a beach is an accumulation of loose sediment, usually sand, deposited by ocean waves over time. A beach extends from the outermost breakers, or breaking waves, to the highest point where the waves reach at high tide. Beyond that zone stand dunes, made of the same sandy sediment but rarely washed by breaking waves.

A beach is composed of three parts:
- **Backshore** the portion of the beach from the berm crest to the dune
- **Foreshore** the portion sloping toward the ocean
- **Nearshore** the portion below water, extending seaward to the outermost breakers

The energy unleashed by waves breaking on the beach changes the beach's profile constantly and rapidly. Different wave rhythms have different effects. In general, high waves with short periods (the amounts of time between waves) cause the beach to erode, and the berm sand is shifted offshore to the bar. Lower waves with longer periods, primarily in summer, move sand from the bar and return it to the berm.

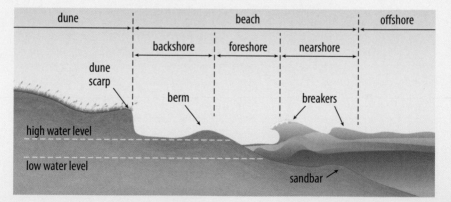

Beachgoers at the water's edge

Sunrise at a Florida beach

Lively surf on the Hawaiian shore

Gentle waves on a sandy beach

THE BEACH ENVIRONMENT

Almost all beaches in the United States are composed of sand, but some are composed of small rocks, including cobbles—smooth rocks up to ten inches in diameter. Most North American beach sand is ground-up granite that rivers have delivered to the ocean, but in some places the sand comes from nearby eroding cliffs.

Some sands include pieces of other rock crystals, shells, lava, or coral. For instance, the white beaches of Florida consist of quartz, seashell, and coral fragments. The black sands of Hawaii are made of lava.

Shaping the Edge

Waves break more than 6,000 times on a beach each day. The waves move sand up the beach in the breakers and down to the water in the backwash, thus reshaping the beach profile with every pound.

Black lava sand

Beach pebbles

Oolitic sand

Quartz crystal sand

Two little girls collect shells on Laguna Beach, California.

The ocean edge is a relatively harsh environment, with saltwater invasions, wind, and waves, but natural beaches support thriving ecosystems in the dunes and the zone between the high- and low-tide boundaries. Healthy sand dunes absorb some of the force of storm waves and block water from washing further inland.

In late winter you're likely to see a narrower beach because winter's stronger storms build larger waves and also push water ashore, making the water higher. Winter waves' backwash carries dune and beach sand into the ocean, where it builds bars. This sand is available to rebuild the beach and dunes in calm weather.

On the dunes you'll find grasses and other plants that hold sand in place and create habitats for invertebrates, including crabs and insects, as well as vertebrates, such as nesting birds and small mammals.

In the intertidal zone you'll find animals that depend on the regular tidal flow and others that come to dine: birds at all times, along with fish and crabs that accompany incoming tides. Rocky beaches, such as in New England and along the Pacific coast, include tide pools—rock basins where water remains at low tide.

FRESHWATER WORLDS

Freshwater features start with springs and streams, congregate in lakes and rivers, and pour out through deltas and bays—thousands of miles of water's edges, each a distinct habitat for wildlife.

Most river-dwelling creatures, plants and animals alike, have evolved in ways that prevent them from being washed downstream. Some have streamlined shapes that resist the flow of water pushing on them. Some stay at the river bottom, where friction slows the water's flow. Others shelter behind or under rocks. Some plants, and even some animals, stay put by attaching to something stationary, such as the river bottom, the side of the bank, or a rock.

From the Lake's Edge

A lake's livelihood depends on its watershed—the source of its water and the material it carries, including pollutants—and its clarity—how deeply light penetrates the water and supports photosynthesis and, thus, the growth of green plants and algae.

Freshwater life-forms concentrate in the littoral zone: the shallow regions near the shore, where light reaches the bottom. Here you'll find submerged plants, rooted plants that stick out of the water, and floating plants. Fish, frogs, turtles, and invertebrates live here.

Clear or Cloudy? Some people may assume that cloudy water in a lake or pond is dirty, but that is not quite true. It is often living matter, not dirt, that clouds the water.
■ If you want to swim and enjoy natural beauty, choose a clear lake or pond.
■ If you want a place that is rich in wildlife and good for fishing, choose a cloudy lake or pond.
Why? Algae can make a lake cloudy, and it can also contribute to a productive food web. Before you fish, consult the locals to be sure the cloudiness is not caused by pollution or harmful algae.

Campers relax by Glacier Lake in the Northwest Territories, Canada.

From the rim of Oregon's Crater Lake, a hiker takes in the view of the azure lake and sheer cliffs.

A kayaker paddles through a saltwater marsh near Core Banks in North Carolina.

MARSHES, SWAMPS, AND BOGS

Throughout North America, distinctive natural communities abound in places where water flows or collects. They can occur far inland or in spots where water flows into the ocean.

■ **Marshes** are low-lying places covered by standing water most of the time. Marsh plants have adapted to saturated soil, and many animal species depend on these environments for food and hiding places.

■ **Swamps** are wooded areas of standing water or saturated soil that may or may not dry up in summer. Unlike marshes, swamps are dominated by trees and shrubs whose root systems are adapted to growing in standing water.

■ **Bogs** are stagnant bodies of fresh water, high in acid content and low in oxygen. They support moss and plants that are specially adapted to this demanding environment.

Prairie Potholes

To see a special kind of marsh, visit the Northern Plains of Montana, South Dakota, North Dakota, and Canada's prairie provinces—southern Alberta, Saskatchewan, and Manitoba. This is where you will find the unique type of wetlands called prairie potholes.

These ancient glacier-carved holes fill with melting snow or rain in the spring. Some dry up by summer, but many stay permanently wet and support submerged and floating plants in the middle and bulrushes and cattails at the edge.

Deemed one of the world's most important wetland regions by the U.S. Environmental Protection Agency, this area hosts half of North America's migratory waterfowl during the summer months. These animals include shorebirds such as the piping plover, the American avocet, and the Wilson's phalarope.

Flat-bottomed boats in a marsh

Pitcher plants in a bog

South Dakota's prairie potholes

ESTUARIES: WHERE WATERS MEET

An estuary is a body of water where salty ocean water and fresh river water meet, creating a unique and productive natural environment. Ocean water measures 35 parts salt per 1,000 parts water, which is equivalent to five ounces of salt in a gallon of water. An estuary contains a mixture of salt water and fresh water called brackish water, which measures anywhere from 0.5 parts per thousand up to the ocean's 35.

■**The San Francisco Bay** and **Puget Sound,** the largest estuaries on North America's West Coast, formed as water filled faults and depressions caused by tectonic plate collisions. Other Pacific coast estuaries, such as Morro Bay, are at the mouths of rivers.

■**The Gulf of St. Lawrence,** through which the waters of the Great Lakes empty into the Atlantic, is the world's largest estuary. It is a semi-enclosed sea flowing through two major straits. Five of Canada's ten provinces border it.

■**The Chesapeake Bay** and **the Albemarle-Pamlico Sound,** both on the Atlantic coast, are North America's second and third largest estuaries, respectively. A

San Francisco Bay

Puget Sound

Gulf of St. Lawrence, Canada

continental shelf slopes out gently underneath these estuaries, thus forming barrier islands—long, low, offshore islands parallel to an ocean coast—and spits, which are fingerlike extensions from the sides of a bay. These formations protect the sound and bay waters from ocean waves, but not from the flow of salt water in and fresh water out through inlets. Rising sea levels filled river valleys to form the Chesapeake Bay.

■ **The Narragansett Bay,** where the Blackstone, Taunton, and Pawtuxet Rivers drain into the Atlantic, represents about 10 percent of the area of the state of Rhode Island.

Estuaries allow salt marshes to form on low-lying areas of sediment protected from an ocean's energetic waves. An estuary's rising and falling tides and small waves gently wash across low shores, carry sediment and nutrients, and support salt-tolerant marsh grasses and other plants. Those plants provide food and havens for a rich web of living things.

Salt marshes provide protective breeding grounds for many saltwater fishes, including several species that spend their adult lives in the oceans. In fact, estuaries are nurseries for 75 percent of the fish caught commercially in the United States.

Chesapeake Bay

Albemarle-Pamlico Sound

Narragansett Bay

Forces of Change

When you look at a beach or at the contour of a bay, you are seeing the results of thousands to billions of years of Earth's sometimes violent geological history. The next time you stand at the water's edge, imagine what geological forces might have shaped it that way.

SHAPING PLANET EARTH

Everything on Earth's surface sits atop a tectonic plate, either one of a dozen or so major plates or one of many minor ones. In some places, two plates are sliding past each other under Earth's crust. In other places, one plate is plunging under another. Where plates are moving apart, magma rises to form new crust.

The plates slide on the asthenosphere, a layer of malleable material 60 to 150 miles beneath Earth's surface. While it is not a liquid, the asthenosphere is easily pushed and deformed—sort of like Silly Putty.

When plates collide, mountains push up. Soon thereafter, moving air and water begin eating away at the mountains in the eons-long process of weathering and erosion. Weathering and erosion shaped today's mountains and, in the process, supplied sand for most beaches along the North American coast.

Moving Water Weathers All

Weathering is the slow transformation of rocks by chemical and physical processes into smaller rocks and, eventually, into sand, silt, or clay. These changes are occurring right now on rocks in moving water, but in most cases they happen much too slowly to notice. The chemical processes include the impact of acids formed by some life-forms that are capable of dissolving rock. The physical processes include the wearing down of stones as they tumble downhill in moving water, or the

A hiker wades in the turquoise waters of Havasu Canyon in Grand Canyon National Park, Arizona.

cracking up of rocks as they freeze and expand. The materials released include trace minerals—nutrients that all forms of life need—which break free from rock and are deposited in soil or sediment as water, wind, and gravity move them. Among these minerals are the salts that streams and rivers carry to the oceans.

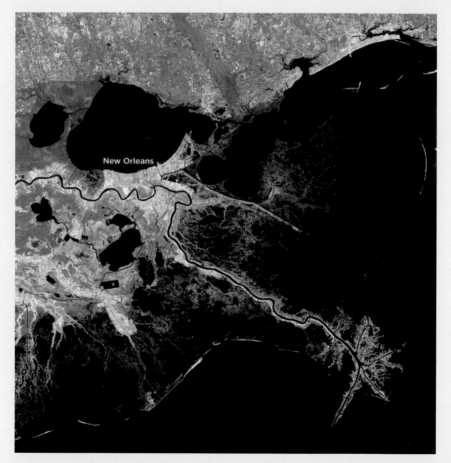

New Orleans

A view from space shows the Mississippi River flowing into the Gulf of Mexico. The Mississippi watershed is the largest in the United States and the third largest in the world, behind the Amazon and the Congo. It covers more than 1.2 million square miles, including all or parts of 31 U.S. states and two Canadian provinces. The watershed covers 41 percent of the area of the 48 contiguous United States.

Ice Shapes Land
Water in the form of ice also helped shape North America. During the ice ages of the last million years, the ice advanced as far south as Kansas and retreated a dozen times. In the process, it gouged out bowls that became northern lakes and crevices that became riverbeds. Since the peak of the last ice age, approximately 18,000 years ago, melting ice has caused global sea levels to rise approximately 400 feet, creating the edges of today's oceans, bays, estuaries, and rivers.

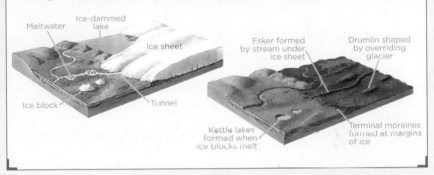

Watersheds Form
Plate movements, volcanoes, mountain building, and moving ice have left us a landscape of large and small bowl-like regions called watersheds, which are areas in which all of the water drains into the same place, whether it falls as rain and snow or comes up from underground and cascades in streams and rivers. Watersheds are separated by divides, sometimes along a mountain ridge and sometimes hardly noticeable.

A watershed is the sum total of land that drains into a single stream, river, or lake. It can be very small, or it can cover a large portion of a continent, depending on the body of water whose watershed you are considering.

Larger watersheds contain multiple smaller watersheds. There are well over 2,000 watersheds in North America. Along the edges of watersheds, from tiny streams and small ponds to larger lakes and rivers, all the way to the marshes and estuaries where the water empties into the ocean, there are worlds to be explored and discovered.

TIDES

In most places along the North American coastlines, you can't spend much time exploring without paying attention to the changing tides. You learn the rhythms and adjust your activities accordingly. Low tide, high tide, low tide: Here is how to understand what's going on.

The tide rises and falls because the moon's gravitational force pulls on Earth's ocean water. Wherever the moon is exerting that pull, the waters bunch up; when the pull weakens, the waters recede.

What Makes Tides?

Here's an extremely simplified explanation of what causes tidal change. Imagine that you are standing on Earth, at Point A, and you can look straight up into the sky and see the moon right now. The moon is exerting its strongest pull on the ocean waters near Point A, and it is high tide where you are standing.

Shoot an imaginary axis from Point A through the center of Earth and out the other side. (Imagine the childhood game of digging a hole to China.) We will call the place where the axis comes out the other side

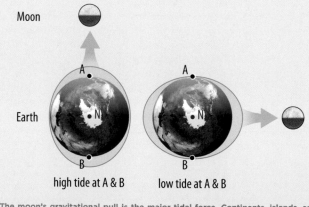

high tide at A & B   low tide at A & B

The moon's gravitational pull is the major tidal force. Continents, islands, and shallows slow water movement, which is why high tide does not coincide with the moon directly overhead.

Point B. It is high tide there right now, too, because centrifugal force is flinging the water away from Earth.

Because of the relative motions of the moon and Earth, it will take 24 hours and 50 minutes until the moon is once again above your head at Point A. It will take half that time—12 hours and 25 minutes—for someone standing at Point B to see the moon overhead. At those two times, high tide will occur at Point A and Point B.

Boats rest at low tide in a canal at Hall's Harbour in Nova Scotia's Bay of Fundy.

Halfway between those two times, low tide will occur at those two places. Thus, on most North American beaches, you can expect 6 hours and 12.5 minutes between the highest and lowest tides of the day.

Neap and Spring Tides

If you return to the same beach again and again, you may have noticed that there are variations in how high a high tide rises and how low a low tide goes. Sometimes that has to do with water far away, but these changes also occur regularly because the sun exerts gravitational force on the waters of Earth's oceans. It combines with the moon's pull to form another tidal cycle—that of the neap and spring tides.

■ **Neap tide** smallest differences between high and low tide water levels when the moon's gravitational pull is at right angles to the sun's

■ **Spring tide** highest high tides and lowest low tides when the moon and sun's gravitational pull combine

We can predict all of these astronomical forces far in advance, thus making it possible to calculate the times and heights of tides years before they occur. But storms, which can be predicted only a few days in advance, can push water into a shore, causing tides to rise higher than they would have on a calm day.

The Geography of Tides

North America has three different kinds of tidal patterns during each lunar day of 24 hours and 50 minutes.

■ **Atlantic coast** semidiurnal tides, with two high and two low tides, where the two lows are of equal height and the two highs are of equal height above or below mean sea level

■ **Gulf of Mexico coast and Alaska's west coast** diurnal tides, with one high and one low

■ **Pacific coast, from southern Alaska through Mexico** semidiurnal mixed tides, with different heights of the two high tides and the two low tides

The Great Lakes are large enough bodies of water that they, too, experience semidiurnal tidal changes, but these changes are so slight that they are hard to notice. For example, Great Lakes spring tides rise less than two inches above normal tide levels in the same location. Wind and rising or falling atmospheric pressure cause even greater water level changes, called seiches.

Each day tides
rise and fall,
weathering an
old pier.

WEATHER AND CLIMATE

Climate—the average of weather conditions over long periods of time—helps determine the natural habitats of water's edge locations. A place's climate depends, first, on its distance from the equator. Prevailing winds, large mountain ranges, and large bodies of water are also important climate drivers.

The Tempering Effect

Since water is slow to warm up and to cool down, large bodies of water have a moderating effect, keeping places warmer in the winter and cooler in the summer. For example, compare Portland, Maine, and Minneapolis, Minnesota, which are on the same latitude but 1,100 miles apart. Portland's proximity to the ocean results in higher lows, lower highs, and an all-around narrower band of change in temperature through the year. Minneapolis gets much colder in the winter and can get much hotter in the summer without the ocean's tempering effect.

Lightning illuminates the purple sky at sunset over Cumberland Island National Seashore in Georgia.

What About Tsunamis? Tsunamis are sudden, huge surges of water that form when an underwater earthquake, landslide, or volcanic eruption creates a vertical movement in the seafloor, thus causing a vertical disturbance at the ocean's surface. As the raised water drops back again, the energy moves away from the source as a tsunami, a series of low, very long, fast-moving waves that can travel hundreds of miles from the source and achieve frightening size and power when they hit a shallow slope. Due to underlying geological activity, the Pacific coast and Hawaii are most prone to tsunamis.

Weather Extremes

The Atlantic is notorious for some of Earth's most spectacular and deadly storms: hurricanes. These storms form over oceans with water temperatures of 80°F or warmer, usually in late summer and early autumn. When an area of low pressure, drawing energy from the warm water, grows into a storm with winds of 74 mph or faster it's called a hurricane. These storms can grow much stronger over the ocean, but many never hit land. Those that hit the East Coast or Gulf of Mexico coast quickly weaken and die but often spread flooding rain far inland.

Florida's Key West suffers the high winds and seaweed-filled storm surge of Hurricane Dennis in 2005.

OCEAN CURRENTS

While they are primarily unseen, ocean currents influence not only the climate but also the movement, the temperature, and sometimes even the color of the water. The Gulf Stream is the best known, and an early map by Benjamin Franklin is an amazingly accurate depiction of this ocean current moving through the Atlantic Ocean toward Great Britain.

Rivers in the Sea

Imagine major ocean currents as rivers in the sea. They carry sea life and nutrients, sometimes from faraway places, and are part of a global system of moving bodies of water that extends deep under the oceans.

When wind and current go in the same direction, wave action is minimum. When wind and current go against each other, a short, stiff wave action develops.
—*ROYCE'S SAILING ILLUSTRATED*

When any ocean life dies, it carries nutrients toward the seafloor as it sinks to be eaten or decomposed. Deep currents collect some of these nutrients and eventually bring them back to the ocean's upper levels near the shore through upwelling.

■**The Gulf Stream** While maps may show a current like the Gulf Stream as one huge swoop, ocean currents are actually much more complicated. The Gulf Stream begins with steady winds from the east that push tropical Atlantic Ocean water toward South America. Some of this water flows north across the Caribbean Sea. The rest flows around the Gulf of Mexico as the Loop Current before squeezing between Florida and Cuba. These streams meet off southeastern Florida to form the Gulf Stream.

The warm Gulf Stream shapes the environment for marine plants and animals on the Atlantic coast, while the cold California Current shapes the environment for marine life along the Pacific.

■ **The California Current** The California current system is not a defined flow like the Gulf Stream but a more complex system extending as many as 600 miles offshore with a northward undercurrent and surface countercurrents.

During the upwelling season of spring and summer, winds blowing from the north combine with Earth's rotational force to push shallow ocean water westward. It gets replaced with cold, nutrient-rich water from deeper down. As the surface current carries this nutrient-rich water along the shore, it enriches the waters and makes locations along the West Coast full of life and extremely productive.

Currents Affect Climate To compare the seasonal influences of the California Current and the Gulf Stream, compare water-temperature ranges at places along the Pacific and Atlantic coastlines:

La Jolla, California	Low (Feb.): 57°	High (Aug.): 68°
Miami Beach, Florida	Low (Jan.): 71°	High (Aug.): 86°

WAVES

Wind creates waves, but the wind you are feeling may not have created the waves you see. Storm winds push on water and stir up waves. The waves grow larger as wind speeds increase, as the wind blows for a longer time, and as the distance along which the wind blows over the water—the fetch—increases.

How Waves Become Swells

Waves sort themselves by size as they leave a storm because waves with a longer wavelength travel faster. They become trains of swells—long, smooth, rounded waves—that transport energy but not water. The water of a swell is moving in an oval orbit, up and down, back and forth. Even on a day when no storm is close, you can see trains of smooth swells moving toward the shore. The arrival and intersection of different trains of swells account for the changes in the patterns and sizes of waves you see.

In the surf zone, waves release their original wind energy when they break. Some energy makes the sounds of breakers, but much goes into the uprush—the water that carries sand up the beach. When the uprush quells, the wave has expended all of its energy, and gravity pulls water back into the ocean as the backwash that swimmers feel as they wade in the surf.

A surfer shoots the curl of a large wave.

Reading the Waves

Watch the water on a beach as a storm approaches, and you will see how the wind shapes the waves. First ripples and then small waves begin forming on top of the swell. The wind takes over and determines the direction in which the waves are heading. When the wind speed reaches 40 miles per hour, it begins blowing the tops off the waves—creating streaks of white and tossing spray into the air. For centuries, sailors have used wave appearance to calculate wind speed.

Riding the Waves

A wave breaks at the beach when the water on top outruns the water on the bottom, which drags on the ocean floor. To bodysurf or bodyboard on a wave, you have to start swimming with the wave just as the top shears away from the bottom. As you surge toward the shore, you use some of the energy that the wave gained from a storm that may have occurred many days ago and hundreds of miles away.

[Life at the Water's Edge]

Some people like to use the nickname Spaceship Earth to express the sense of how this planet supports its inhabitants. Any spaceship, including Earth itself, needs to supply oxygen, food, and water for those aboard. When you visit a water's edge, consider how the environment provides for its life-forms, which interact in a complex web of nature.

LIFE'S ESSENTIALS

Water is vital for all life. Since so many substances readily dissolve in water, it transports essential ingredients among all the cells of a plant or animal.

Animals that live on land obtain the oxygen they need from the air. Marine mammals come to the water's surface to breathe air. Marine animals with gills use oxygen that is in the water.

Plant photosynthesis releases oxygen into Earth's air and water. In the simplest of terms, with the help of energy from the sun, plants take in carbon dioxide (CO_2) and water, break them down, and recombine them as sugar (a carbon-based molecule) and oxygen (O_2).

Photosynthesis Happens

Photosynthesis occurs in plants large and small, aboveground and underwater. It takes place in algae and in microorganisms such as plankton and bacteria. Cyanobacteria (also called blue-green algae), one of the most common photosynthetic producers, live both on land and in the water. Even though they are microscopic bacteria, they are often visible as clusters of biofilm.

A shortage of oxygen can kill fish and other animals living in the water. Since warm water holds less oxygen than cold water does, fish kills are more common in summer

A large bull elk drinks from a woodland stream in western North America.

than in winter. A large "bloom" of algae, caused by extra nutrients such as nitrogen and phosphorous that washed into the water, can deplete oxygen. When the algae die, they begin to decompose, and the organisms responsible for decomposition take oxygen from the water, thus leaving too little for the fish and other marine life present. Creatures that can't swim to water with more oxygen die.

That Certain Glow Many life-forms in the ocean display bioluminescence. In other words, they glow. Most of these organisms dwell deep underwater and are visible only to research divers and other animals, but some glisten in shallow water and are visible from a beach or dock in just the right conditions—particularly on a moonlit night, as the waves beat against sand or piling. When you see this glow in the water, most likely you are looking at dinoflagellates, single-celled algae that thrive in seawater and give off a neonlike light when disturbed.

Succulents thrive seaside in Monterey Bay, California.

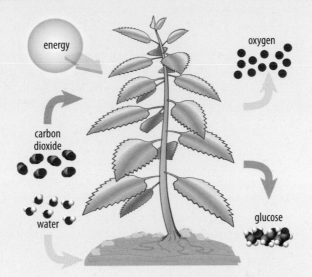

energy

oxygen

carbon dioxide

water

glucose

Photosynthesis is the process by which plants derive energy from sun, air, and soil.

PLANTS AT THE WATER'S EDGE

The plants you see along the water's edge offer glimpses into how living things have evolved to thrive in different environments.

Some plants, such as eelgrass or rockweed, germinate, grow, and reproduce entirely underwater. Others, such as water lilies, grow halfway in, halfway out, with leaves that float on top of the water. Many sink roots into the sediment at the edge of the water but shoot leaves and flowers up into the air.

All the water's edge and underwater plants have evolved somewhat differently from land-dwelling plants, as they have more direct access to the water that is so essential to all plants for building cells and transporting nutrients. These plants' roots do the work of anchoring and transporting nutrients, but they do not have to grow deep or broadly, as do some roots that sink into dry soil.

Most land plants develop stems strong enough to hold up the leaves. The ultimate examples are trees, whose trunks strengthen over years and support branches with their loads of leaves. Water-dwelling plants need not develop that same rigidity, since they float upward or lie on the surface of a lake or pond. Instead, they depend upon water for physical structure; thus, if the water they are growing in dries up, the stems won't hold up the leaves.

The sun's rays are the ultimate source of almost every motion which takes place on the surface of the earth. By their heat are produced all winds . . . By their vivifying action vegetables are elaborated from inorganic matter.
—SIR JOHN HERSCHEL, *OUTLINE OF ASTRONOMY* (1864)

The Challenge of Life on Land
Land-dwelling plants have several ways to bring in and keep in essential moisture.

■ The cell structures of many roots, stems, and leaves successfully hold water in, and the plant's pores open and release water only when necessary.
■ Tubelike structures, or vessels, transport water and nutrients in a complex exchange essential to plant life and photosynthesis.
■ Pores in the leaves, called stomata, release oxygen and water vapor.

Water molecules are strongly attracted to one another. When water molecules evaporate into the air as water vapor—called evapotranspiration—their attraction pulls the water molecules behind them in the tube toward the stoma.

In addition to allowing water to exit the plant, the stomata serve as entrances for carbon dioxide, which, through photosynthesis, the plant will combine with water that reaches the leaves to produce nourishing sugars.

Cardinal flower

Spike moss

Blue flag iris

Mosses on rocks and a dead log

Yellow water lily

Field horsetail

Skunk cabbage flowers

Cattail

Swamp saw grass

VARIETIES OF LIFE

Many of the creatures that you will get to know as you explore the water's edge are classified as invertebrates. The animals in this diverse group have one thing in common: they do not have backbones. The clams, oysters, mussels, lobsters, and crabs that we feast on are all invertebrates. That mosquito buzzing around your head is an invertebrate. Sea urchins and jellyfish, crawfish and snails . . . from tiny worms to giant squid, invertebrates account for approximately 97 percent of all known animal species.

How Do Invertebrates Do It?
The varied world of invertebrates displays many differ-ent body plans, and it can be fascinating to learn how each group of species takes in food and nutrients to fuel growth and movement. In some, such as sea stars and

Crayfish

Whitetail dragonfly

Purple sea urchin

Moon jellyfish

Bivalve mollusk

American lobster

Snow geese fill the sky during their spring migration at Squaw Creek National Wildlife Refuge, Missouri.

sponges, water circulates through the body and carries food and oxygen to the cells. In others, including crabs and earthworms, a primitive heart circulates blood. Many invertebrates, including corals and jellyfish, have simple digestive systems with one end opening to take in food and another to excrete waste.

Smart Squid Many invertebrates have rather primitive nervous systems. But the Humboldt squid, which lives in the Pacific off the coast of the Americas, hunts fish in groups in ways that require cooperation and communication among individuals. Observations of these creatures in their natural habitats, as well as laboratory studies, have shown that cuttlefish, squid, and octopuses have intellectual abilities equal to those of some vertebrates.

ANIMALS AT THE WATER'S EDGE

Whether it lives on land or in water, any animal needs to obtain food and oxygen, extract the food's nutrients, and deliver oxygen and nutrients to its cells. The chemical reaction of oxygen and glucose from food produces energy, carbon dioxide, and water.

Lungs for Breathing

Animals can live for a while without food, but a lack of oxygen is quickly fatal. All mammals, birds, and other relatively complex land animals use lungs to take oxygen from the air and to get rid of carbon dioxide. Lungs consist of small air sacs with extremely thin walls adjacent to tiny blood vessels. Oxygen molecules move from the air into the blood, while carbon dioxide moves from the blood out into the air as we breathe. Red hemoglobin in the blood carries oxygen and carbon dioxide.

Spadefoot toads breed in a temporary pool in Maryland.

Northern elephant seal

Brown pelican

Gills for Respiration

Fish and other animals that live in water use gills—thin layers of tissue containing small blood vessels—to breathe underwater. As water passes over the gills, oxygen enters the blood vessels, and carbon dioxide is exhaled into the water.

Amphibians such as frogs and toads, which begin life as fishlike tadpoles, have gills but acquire primitive lungs as they take on their adult forms. Adult amphibians also absorb oxygen through their skin. Ghost crabs, which have gills but live in burrows in beach sand, need to scuttle to the water periodically to wet their gills.

Very small animals respire, or take in oxygen, without lungs. Their cells are close enough to the animal's outer layer for oxygen to penetrate directly. Even insects have networks of small tubes that bring oxygen to all of their tissues.

Varieties of Eating

As animals became larger and more complex, they evolved a variety of methods to obtain nutrients and oxygen, to keep warm, and to reproduce. More complex animals have digestive systems to process what they eat, a circulatory system with blood vessels and a heart to pump blood that carries oxygen and nutrients to the cells, and a central nervous system to coordinate it all. One of the fascinations of observing animals along the water's edge is learning how they have evolved to meet these basic needs, whether they live in the water, on land, or in both places.

For example, during a day at the beach you might see a jellyfish in the shallow water. It absorbs oxygen through its skin and uses the central cavity of its sacklike body both to process food and to distribute nutrients. It does not need a central nervous system.

But out in the ocean, a whale's circulatory system, like those of other mammals, supplies oxygen and nutrients to its cells, takes away waste, and helps regulate body temperature. When the whale dives deep underwater, the circulatory system slows the heart rate and restricts circulation to the extremities, thus keeping the whale's heart, brain, and lungs well supplied with oxygen. The whale's central nervous system coordinates these complex activities and supports the animal's social life.

WEBS OF LIFE

From the microscopic to the majestic, all of the plants and animals that you see along the water's edge (or anywhere else) are part of an interlocking food web integral to that environment. Here are the players:

■**Producers** plants or algae that use photosynthesis to produce the various compounds that comprise them

■**Consumers** animals that eat plants, algae, or other animals to supply energy for growth and movement and material for their cells and tissues. Herbivores eat only plants; carnivores eat other animals.

■**Scavengers and decomposers** animals that recycle the nutrients from dead plants and animals. Scavengers eat dead plants and animals. Decomposers, which include fungi and microorganisms, break down animal waste and dead plants and animals.

A food web is made of simple links, but it is more than just a one-dimensional chain from plant to herbivore to carnivore. Consider the blue crabs, native to the Atlantic waters along the North American coastline. The crabs are omnivores because they eat both plants and animals. They are also scavengers because they eat dead plants and animals. Blue crabs are eaten by other animals—skates, fish, birds, bigger blue crabs, and people. Even sea stars (commonly called starfish), which are relatively simple invertebrates, prey on blue crabs during winter, when they find them dead or dormant. Blue crabs illustrate one example of the many interlocking food webs to be found at the water's edge.

Masses of white pelicans feed in the nutrient-rich waters of the Mississippi Delta.

FIELD GUIDE

Gather a shell from the strewn beach
And listen at its lips: they sigh
The same desire and mystery,
The echo of the whole sea's speech.
—DANTE GABRIEL ROSSETTI, "THE SEA LIMITS"

[**1**]

BEACHCOMBING

**Large conch shells are among the most
beautiful items that beachcombers may find.**

Beachcombing Basics

Almost every beach visitor does a little beachcombing along the ocean's edge to see what the tide has washed in. It is a relaxing escape from everyday concerns. Other people take their beachcombing seriously, making a hobby or even a business of collecting what they like among the many objects they find. Whether you are casual or serious, beachcombing helps you connect with the ocean and some of its creatures.

Rows of vegetation along the upper beach help to hold low dunes in place.

Start With the Wrack Line

Wrack refers to several species of living seaweed, as well as dead vegetation of all kinds, that are washed up by waves and often are concentrated along the high-tide line, which is called the wrack line. This long row of debris stretching along the beach almost always offers fascinating discoveries—for example, the seaweed called sargassum, which contains little brown balls. The balls keep the sargassum afloat as it circles the Sargasso Sea, a huge eddy in the middle of the Atlantic. Pieces of sargassum, including their little floats, break off and drift to beaches along the Atlantic and the Gulf of Mexico. In wrack lines you can find many of the items that are described in this chapter. In fact, wrack lines sometimes contain more living things than any other part of a beach. Walk along the line slowly, look carefully, and you may find many creatures that you have never seen before. Wrack offers protection, nutrients, and moisture to an entire food web that includes tiny shrimplike creatures, mole crabs, ghost crabs, and shorebirds such as terns that are foraging for a meal.

safety tip

Do not pick up or prod unrecognized man-made items such as metal canisters; unexploded bombs or mines sometimes wash ashore.

Seeds and nutrients in wrack washed onto the back beach may help plants become established to stabilize new dunes.

Rules for Beachcombing

Before collecting any seashells, be aware that there may be strict government regulations to follow or private property to respect. In many locations you are not permitted to take whatever strikes your fancy. In some areas, it may be forbidden to collect seashells containing live animals and even to take empty shells. Always check for possible regulations because illegal collecting may result in substantial fines. Two examples: 1. Pay attention to strict regulations banning the collection of abalone, which are large, edible sea snails, along the California coast. 2. Florida prohibits collection of beautiful queen conchs, which have edible meat.

Even where collecting mollusks is legal, you should not interfere with nature's food web by taking any living mollusk. Taking items besides living creatures may also be prohibited. For example, the National Park Service forbids removing sea glass or pottery from Spectacle Island in Boston Harbor, a prime site for finding sea glass.

It is especially important to learn the rules before beachcombing at national wildlife refuges, national seashores, and state parks—but wildlife regulations may also apply to private property. There is another good reason not to collect shells containing living animals: When the creature dies and begins to decay, its odor may quickly become unbearable, perhaps like that of rotting fish.

Sea Glass and Plastic

Many beachcombers collect pieces of glass that waves, sand, and water have tumbled and smoothed over many years, creating unique shapes, colors, and textures. Some collectors use them to make sea glass jewelry or art objects. The North American Sea Glass Association links collectors with its newsletter and collecting conventions. As plastic containers increasingly replace glass ones, sea glass is getting harder to find. Most beachcombers view the large amount of plastic waste on beaches as unsightly garbage.

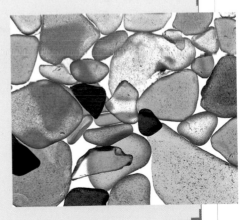

Common Shells

Most shells that you are likely to discover on a beach are (or were) mollusks' external skeletons, which protect the animals' soft, fleshy bodies. All species of mollusks are important parts of marine food webs. If a shell is empty, another creature has eaten the animal, or it has died and rotted. Twin shells attached by a hinge, such as clams, are called bivalves. One-piece shells, such as snails, are called univalves.

A scallop is held open to display the tasty white muscle meat inside.

Scallops

family Pectinidae
L up to 8″ **F** **S**

appearance These fan-shaped bivalves with low ridges and fluted edges come in a variety of beautiful colors. Usually they are bright brown or reddish when they have not been bleached by the sun or worn by the surf.

origins The larvae of scallops drift as tiny plankton. Adults propel themselves along the sea bottom by opening and closing their shells. Some species use a cordlike structure to attach themselves to a substrate; others cement their shells to a substrate; yet others use a body part called a foot to burrow into the sand.

range Oceans and bays worldwide.

key sites Ocean and bay beaches.

Bay scallops

Ark shells

Ark Shells

family Arcidae
L up to 3" **S**

appearance The 200 species of arks come in various sizes, and their shells are either elongated or square. When alive, these clams have dark, velvety, furlike covers, but when found on beaches, the covering is often eroded by the surf and the shell is bleached white or beige by the sun. These animals are named for their shells' perceived resemblance to Noah's ark.

origins Arks are filter feeders that dine on plankton and algae. They use their strong feet to anchor themselves to sand or mud on the seafloor. They are the only bivalves with red blood.

range Tropical and temperate oceans worldwide; common on Atlantic and Gulf coasts

key sites Beaches, especially after storms.

did you know ? **tiny bubbles** popping up at the surface of wet sand may come from coquina clams below? A half inch long, these clams have many colors and patterns. Holes in the sand indicate burrows of larger clams.

Cockles

family Cardiidae
L up to 8" **S**

appearance More than 200 species come in various colors and are often peach inside. When the twin shells are connected, most are heart-shaped when viewed from the side. Many also have ridges called ribs, which radiate from the hinges, although egg cockles' shells are smooth.

origins Cockles live in shallow tidal salt water off beaches and in estuaries. Normally, adults remain in one spot and feed on microscopic plankton by filtering water. These mollusks sometimes move along the bottom by using their feet to spring in two- to three-foot hops. Because of cockleshells' heart shape, many authors have used them in historical and literary imagery. Examples range from the nursery rhyme "Mary, Mary, Quite Contrary" ("with silver bells and cockle shells") to the Roman Catholic religious emblem of St. James, the patron saint of Spain.

range Worldwide.

key sites Cockleshells often wash up on ocean and bay beaches.

Cockleshells are often tinged with peach.

A collection of whelk shells, some broken

Whelks

family Buccinidae
L up to 2-16" **S**

appearance Whelks' spiral shells taper to a point at one end and have a long channel at the other end. The surface is usually smooth, with whorls of sculptured, coil-shaped indentations. The indentations begin at the tiny point and increase in size as the animal grows to adulthood. If you look at a whelk shell from the front, the hole called the aperture, where the animal emerges, may be on the right side or the left side. These are known as right-handed or left-handed shells.

origins There are more than 1,500 species, a few living in freshwater. Whelks feed on worms as well as other mollusks. Most species drill through bivalve shells to reach the meat inside. Females lay spongy capsules with hundreds of eggs, few of which hatch. Many of the rest wash up on beaches.

range Worldwide, intertidal zone to deep ocean.

key sites Beaches along oceans from the tropics to the polar regions.

Murex snails produce dye.

Purple Dye of Royalty

More than 2,000 years ago the Phoenicians used mucus secretions from two marine snails—*Murex brandaris,* or spiny dye-murex, and *Buccinum lapillus*—to produce the original (and very expensive) royal purple dye. It was named Tyrian purple after a center of the ancient dye industry, the city of Tyre in today's Lebanon. The ancients prized it because the color did not fade; it became brighter and more intense over time as the dyed garments were washed and worn in sunlight. The Old Testament, the Greek poet Homer, and Greek and Roman historians describe the dye and the cost of producing it. The Roman Empire allowed only the emperor and other high-ranking people to wear such garments—thus the term *royal purple.*

look for colorful **apple murex** shells early in the day because they quickly fade in sunlight when they are out of the water and dry.

Apple Murex

Phyllonotus pomum
L up to 2-4.5″ **S**

appearance The shell is heavy, with thick and thin ribs, sharp spikes, and a large aperture where the snail emerges. Colors are variegated shades of white, tan, and brown.

origins This murex feeds on other mollusks, barnacles, worms, and invertebrates. It bores holes in mollusks to reach the meat inside. The female lays capsules containing 80 to 200 eggs. There is no larval stage; young emerge as crawling individuals.

range North Carolina to Brazil in the Atlantic, Caribbean Sea, and Gulf of Mexico.

key sites Best on sheltered beaches in Florida, where high surf does not break off the spines.

Spiky apple murex shells

Conch shells are always a delight to find.

Conchs

families Strombidae, Melongenidae
L up to 12″ **S**

appearance The name *conch* refers to various species of large sea snails, generally having conical shells, often flared out and with blunt spikes. Colors vary by species; in the ocean, shells are often covered by algae and debris.

origins The animal hatches in shallow water, and larvae drift with currents for a month before settling on the seafloor to begin growing. The spiral shell has spiky bumps that become blunt with age. The snail moves into shallow water to lay thousands of eggs. Queen conch meat is prized for making tasty conch chowder, and it is declining in much of its range due to overharvesting.

range Oceans worldwide. Six species are found in the Caribbean area from Bermuda and southern Florida to southern Mexico, Venezuela, and northern Brazil.

key sites Shells are sometimes found on shallow-water reefs and sea-grass meadows, but usually on beaches.

Snail Shells

Many collectible "seashells" are marine snails such as murexes or conchs, which are among the estimated 200,000 snail species found in oceans, in fresh water, along water's edges, and on dry land (even in deserts!). They range from as tiny as a grain of sand to as large as a two-foot-long Florida horse conch. Extremely diverse, they can be plain or dazzlingly colored and patterned; smooth, ridged, or spiked; round and flat; or tall and narrow.

Some snails live away from the sea, such as this colorful tree snail.

Small Snails
L up to 2" **F** **S**

The following small species are found alive in tide pools or dead on beaches:

Common periwinkles (up to 2") are edible species that live on rocks along both the Pacific and the Atlantic coasts. They vary from grayish brown to brown with light bands.

Violet sea snails (1–1.5") float on excreted mucus bubbles. Tiny bubbles in wrack could lead you to these beautiful snails, but take care: Sometimes they hitch a ride in stinging jellyfish tentacles.

Black turbans (1–2") are common in Pacific coast tide pools. They have a black-purple shell and often live in colonies.

Nerites (0.25–2") These tiny animals include such colorful—and colorfully named—species as the bleeding tooth and the zebra nerite. They live on rocks between the tide lines.

did you know?

a live animal could be in a snail shell you find? Put the shell into a bucket of seawater for a few minutes. An animal may extend out of the shell. Return any living mollusks to the sea.

Frilled dogwinkles

Frilled Dogwinkle

Nucella lamellosa
L up to 4" S

appearance Like other whelks, dogwinkles have spiral cones. Some are frilled, and some are smooth. Frilled shells are found in the northern part of the range. The shells are white, light brown, gray, or orange, and some have colored bands.

origins The frilled dogwinkle is a major predator of acorn barnacles. The female can lay 1,000 eggs in winter or early spring; the young hatch in a month or two and crawl away. Only about one percent of the young live as long as a year. The dogwinkle can drill a hole in its prey, may inject poison, and uses its tongue to scrape out meat.

range Along Pacific coast, from Bering Strait in Alaska to central California.

key sites Intertidal zone or just below the low-tide line on rocks and in crevices.

look for **yellow egg cases** of frilled dogwinkles attached to rocks near the low-tide line on Pacific coast beaches in winter and early spring.

Bleeding Tooth

Nerita peloronta
L up to 1.5" S

appearance The outside of the shell is grayish or yellowish, with irregular black and red zigzag markings. The underside, rimming the aperture, is flattened and has an orange to scarlet area edged by several large and small "teeth" (decorative, not used for eating), for which the animal is aptly named.

origins This species eats various kinds of algae, diatoms, and cyanobacteria that coat rocks like a film. Unlike many snails, it reproduces sexually.

range Common in Florida, Bahamas, West Indies.

key sites Rocks and tide pools in the intertidal zone, as well as on reef flats. Related species live on mangrove roots, sea grasses, and mudflats.

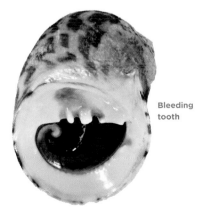

Bleeding tooth

Limpets & Oddities

Keyhole limpet

Limpet is a name used for a variety of mollusks with shells shaped like a dish, a cone, or a pan instead of being coiled like a snail's shell. This diverse group of mollusks includes keyhole limpets, which have a hole at their apex, and true limpets, which are peaked and lack an apex hole. All species have a muscular foot that enables them to adhere to rocks. Species that live on rocks exposed to harsh wave action have the flattest shells; those in calm or deep water have the highest shells. Some distantly related species called limpets live in freshwater lakes.

True limpets

True Limpets

family Patellidae
L most up to 8" **S**

appearance The true limpets come in diverse colors and patterns on cone-shaped shells but are often bleached dull gray when found on beaches. Alive, they may be covered by microscopic algae that camouflage them against rocks.

origins Limpets graze on algae-covered rocks. Their yearly spawn is triggered by rough seas that disperse the eggs and sperm. Larvae float around at first. They mature into males in nine months and into females in two years. After grazing at high tide, they follow their mucus tracks back to their home base as the tide goes out.

range Worldwide, from shallow water to deep ocean at hydrothermal vents.

did you know ?

a true limpet's foot acts like a suction cup? It is combined with a glue that the animal secretes to hold it firmly on a rock despite pounding by waves. Its tight hold on the rock also deters predators.

Worm shells form a strange and unsightly tangle of tubes.

Worm Shell

family Vermetidae
various lengths **S**

appearance These irregular, tube-shaped shells are cemented onto a hard surface or onto one another, sometimes in masses that form reefs. When alive they are dark, but they are most often found bleached white when dead.

origins The shells are coiled in the free-moving larval stage, but then they grow into intertwined colonies. Worm shells feed by releasing into the water nets of sticky mucus, which they pull back with tiny food particles attached.

range Worldwide in tropical- and temperate-zone waters.

Atlantic Auger

Terebra dislocata
L 2.25" **S**

appearance Long, pointed spire, whorls wind around 20 to 25 low ridges from top to bottom, varied exterior with bands of pale gray, pinkish brown or orange-brown, slightly corrugated appearance. Name refers to similarity to tool used to drill holes.

origins Unlike other augers does not poison prey, not a threat to shell collectors, no larval stage of life, hatched young crawl away.

range Common in Atlantic Ocean sounds and offshore on shallow sand flats from Maine to Brazil, around the Gulf of Mexico, in the Caribbean Sea.

Atlantic auger

A carrier shell with its collection of add-ons

Carrier Shells
The first time you see a carrier shell, you might think it is a practical joke by someone who glued objects to the shell. Actually, this shell uses its foot to lift and cement pieces of other shells and rocks to its upper shell, and it adds more as it grows. The aggregation disguises the shell as a little pile of debris. Carrier shells, which are one to six inches in diameter, live in tropical and subtropical oceans.

Cowries & Olives

Cowries are members of a worldwide mollusk family that contains hundreds of species. Beachcombers like to collect them for their brightly colored shells, which often look as though they have been polished. They are snails, though they do not have coiled shells. Olives, which collectors also enjoy finding, are members of another large family that spans the globe. Many olive species resemble one another closely and can be difficult to identify.

A chestnut cowry scours the seafloor for a meal off southern California.

Cowries

family Cypraeidae
L up to 6" **S**

appearance The adult shell is rounded, with a long, narrow opening on the underside. When the animal is moving, the shell is usually covered with a mantle of a different color and pattern. The snail secretes a pearl-like substance that keeps its shell shiny.

origins Cowries feed on algae and tiny sea life on reefs and the sea bottom. They feed at night and hide during the day. The first shell is a narrow spiral, later covered with the adult outer shell. Full colors and patterns appear only in adults.

range Tropical, subtropical, and temperate oceans.

key sites The chestnut cowry is common near the low-tide line in southern California; other species live at various depths.

A cowrie shell

Tent olive

Olives

family Olividae
L 1-5" **S**

appearance The shells, which resemble olives, are cylindrical, smooth, and shiny. They occur in many patterns, and the shells of some species have numerous fine wrinkles.

origins The 400 species worldwide eat snails, bivalves, and small crustaceans, and they are also scavengers. Most burrow into sand. An olive finds prey via its sense of smell, engulfs the prey with its large foot, and smothers it with slime.

range Worldwide; 25 species occur on the shores of North America.

key sites Sandy shores.

Cowries in History
People around the world have used cowry shells as money for centuries. A Pacific and Indian Ocean species, *Monetaria moneta*, the money cowry, is used in Africa, Asia, and on Pacific islands. Archaeologists found hundreds of cowries in a 1760s trash dump in Yorktown, Virginia. A slave trader had used them to buy slaves from Africans. Cowry shells and gold replicas appear in the tombs of Egyptian pharaohs dating back 4,000 years. People of other cultures have used cowries as decorations for clothing and other objects, as well as in religious and other rituals.

Atlantic Slipper Shell

Crepidula fornicata
L up to 2" **S**

appearance Various species' shells are highly or slightly arched, rounded, or flat. The Atlantic species is about average in height. The colors are yellow, creamy, or brownish, often flecked with reddish brown. A white "deck" on the inside of the shell gives it the appearance of a boat or a slipper.

origins These animals feed on plankton and small detritus. They live on rocks, on other shells, and even on the backs of horseshoe crabs—often in stacks, which ease reproduction. During spawning, the animal changes from male to female and back again.

range Atlantic coast and Gulf Coast, from Canada to Texas.

key sites Beaches.

Atlantic slipper shell

ON LOCATION

Sanibel Island Florida

- Shell collecting
- Wildlife refuges
- Bike and walking trails
- No mid- or high-rises

At times Sanibel Island beaches seem to have more shells than sand.

Sanibel Island and its sister, Captiva Island, are among the rare Florida beaches where nature rules. Sanibel is famous as one of the world's best places for finding seashells. Dr. Beach explains that waves and tides rip seashells from an offshore bank and push them ashore. The gently sloping ocean bottom reduces rough surf that smashes shells elsewhere.

Shell collecting is only one reason to visit Sanibel and Captiva. Wildlife refuges cover more than half of Sanibel Island, which is approximately 12 miles long and 5 miles across at the widest. For a break from the beach, you can explore the island's varied inland and bay habitats. The largest refuge is the 5,000-acre J. N. "Ding" Darling National Wildlife Refuge. It is named for Darling (1876–1962), a pioneering protector of wildlife who served in the 1930s as the first head of what is now the U.S. Fish and Wildlife Service. The refuge includes undeveloped important southwestern Florida coastal habitats, including more than 6,400 acres of mangrove forest, submerged sea-grass beds, cordgrass marshes, and hammocks, which are strands of hardwood trees on natural rises in marshes. More than 220 species of birds either live on the island all year or stop there during their fall and spring migrations. Many visitors find birds are a bigger attraction than shells.

Terns gather in large numbers on a Sanibel Island beach.

Nature Without Roughing It

Sanibel and Captiva do not have buildings higher than three stories, huge paved parking lots, megamalls, expressways, or manicured beaches. These islands are natural, with wrack, shells, and driftwood.

Instead of highways, Sanibel has more than 20 miles of paved bicycling and walking trails, two-lane roads with a 35-mph speed limit, and no traffic lights. If you are a nature novice, several companies offer guided tram, boat, and walking tours, or guided fishing trips, to introduce you to the island's wildlife. The Bailey-Matthews Shell Museum's exhibits and talks are good resources for learning about shells.

After a day of looking for shells, biking, wildlife watching, or lounging on the beach, you could enjoy a sight that bright cities and suburbs have taken from most of us: a sky full of stars, with the sparkling band of the Milky Way stretching from horizon to horizon. Thanks to the folks who enforce lighting restrictions, you can settle down in the evening to enjoy nature's starlight.

Paddling a kayak or canoe is a good way to explore the water's edges around Sanibel and Captiva Islands. A small, quiet boat offers access to parts of the Ding Darling Wildlife Refuge and other island areas you won't see from walking trails. Paddling also allows you to quietly approach wildlife you want to view or photograph. Several operators offer kayak or canoe rentals, and some offer tours guided by naturalists.

The islands have hotels, ranging from modest to luxury, plus small rental cottages, luxury resorts, and condos. When it is time to eat, you can choose from takeout delis or restaurants that offer impressive wine lists.

All in all, Sanibel and Captiva offer a taste of Florida's coastal world before developers began destroying the natural environment in the 1920s.

Coral reefs in shallow seas are often exposed above water at low tide.

Corals & Sponges

Corals wash onto beaches only where coral reefs are offshore. The sands of some beaches in Florida and Hawaii consist of ground-up coral, but large clumps of coral are rare on most beaches, except after a hurricane rips into an offshore reef. Nevertheless, some places in Hawaii and the U.S. Virgin Islands, where reefs are close to shore in shallow water, have coral rubble beaches, with pieces of many types of coral. Beachcombers may find many types of sponges, which live anywhere from brackish marshes to deep ocean bottoms. As with corals, storms can deliver more and larger sponges than those usually found.

Corals

wide variety of species
L varies greatly **S**

appearance Coral rubble comes in a wide variety of sizes, shapes, and colors. Wave battering usually destroys the coral's original shape and leaves bleached white pieces. On rare occasions, the colors may be preserved.

origins Colonies of many individual polyps secrete calcium carbonate to build hard skeletons in different shapes depending on the species. These diverse types cluster together and grow larger over decades or centuries to form reefs.

range Tropical and warm oceans worldwide, including South Florida's east coast, Hawaii, Puerto Rico, and the U.S. Virgin Islands.

key sites Best variety and numbers along the Florida Keys.

A beautiful coral fragment

Boats in shallow water may damage coral by scraping or breaking it.

Endangered Reefs Coral
reefs have survived a host of natural threats through evolved defenses, such as the ability to recover from storm damage. Now, human activities are directly damaging coral or weakening its responses to disease and other dangers. Many scientists estimate that if current trends continue, 60 percent of the world's coral reefs will die by 2050. Increasing ocean water temperatures, which may be caused at least in part by human activities, are a major threat. Increased pollution from coastal development, poorly treated sewage, and agricultural runoff weaken reefs' defenses against disease.

Coral reefs support the most diverse biological community of any marine habitat and support many species of beautiful tropical fish. Many reefs are popular with divers, who collect fish and coral for themselves or for sale to aquariums. Large pieces of coral are frequently damaged when boats strike reefs in shallow water. Government protection in marine sanctuaries and education of the public by conservation organizations are important elements of maintaining healthy coral, but the task is difficult because corals face so many different threats.

Sponges
wide variety of species
L up to 6' **S**

appearance As with coral, different species come in a variety of colors and shapes. Most are no bigger than hand size.

origins Unlike other marine invertebrates, sponges have no internal organs. Their bodies consist of a network of chambers and passages that transport water with food (bacteria and tiny particles) to all parts of the animal. This structure has been described as a group of single cells that work together.

range Most of the 5,000 to 10,000 species live in temperate-zone oceans at many different depths. A few species live in fresh water.

key sites Sponges can wash up on any beach, especially in storms.

A sponge washed onto a beach

safety tip

Be careful when handling a sponge. A few sponges produce toxins that can sting, cause itching, or otherwise irritate your skin.

Beach Castings

One pleasure of beachcombing is encountering strange creatures that have washed onto the beach. Sea urchins and their close cousins, sea biscuits and sand dollars, are popular for collecting. Sand dollars are rounded and have five sides. Sea urchins, when alive on the ocean floor, are covered with long, strong spines. Living sand dollars and sea biscuits, which burrow under the sea bottom, have short, velvety spines or velvety, spineless surfaces.

Sand dollars often have attractive star-shaped markings.

Sand Dollars

many biological families
L up to 6" **S**

appearance Living animals are covered by a skin of velvety spines coated with very small hairs. They come in a variety of colors, including blue, green, violet, and purple, but dead sand dollars found on beaches are nearly always bleached white and have lost their skin.

origins Sand dollars use their spines to dig into the seafloor. They feed on crustacean larvae, small copepods, diatoms, algae, and detritus.

range Widespread in temperate and tropical oceans on sandy or muddy seafloors, from the intertidal zone to great depths.

key sites Sandy beaches on Atlantic, Gulf of Mexico, and Pacific coasts, especially after storms wash them ashore.

look for **tiny, fuzzy hairs** that cover the surface of a sand dollar. If these hairs move, you will know the animal has been washed onto the shore but is still alive. If so, you should return it to the sea.

A sea biscuit on a beach

safety tip

Sea urchins' spines can penetrate your skin and break off. Most are non-toxic, but some release venom that causes stinging, numbness, paralysis.

Sea Urchins

many biological families
L most up to 5" **S**

appearance Sea urchins have globular shells with long spines radiating from their bodies. Common colors include black, dull shades of green, olive, purple, and red. When found on beaches, the spines are often worn away.

origins These slow-moving creatures feed on algae, sea cucumbers, and other invertebrates. Among the spines are rows of tiny tube feet with suckers that help with locomotion, capturing food, and holding on to the seafloor.

range Worldwide (more than 700 species).

key sites washed ashore from the western Atlantic and Caribbean ocean bottoms.

Spines are lost after the surf washes a dead sea urchin onto the beach.

Sea Biscuits

many biological families
L up to 10" **S**

appearance These creatures are closely related to sand dollars. They have oval bodies, and some have five symmetrical oval "petals" that extend over all or part of the upper surface. Generally, they are not as flat as sand dollars, and they may have short spires of various colors.

origins Adults live semiburied in coarse, biogenic sand of coastal waters, where they feed on organic matter in sediment. The sea biscuits use accessory spires around the mouth to collect these edible sand particles and then crush them with five calcium carbonate "teeth."

range Sea biscuits are fairly common on the Atlantic coast from North Carolina to the Caribbean Sea to São Paulo, Brazil; also in the Gulf of Mexico.

key sites Sea biscuits can be washed ashore on any beach.

The large egg case of a shark has flanges that help to anchor it in crevices.

Egg Cases

L up to 4" **F** **S**

appearance Egg cases of sharks and skates (a type of ray) are usually dry, black, and empty of eggs when washed up on the beach. Some are rectangular, with "horns" on the corners; others are cylindrical with spiral flanges. Big skates' cases on the Pacific coast are up to 12 inches long and 7 inches wide. A moist case with no holes could hold a live embryo. Egg cases of large snails called whelks are long, chainlike structures, and few people would guess that they once held eggs.

origins Some sharks and skates produce embryos in sacs; eggs take 3 to 15 months to gestate and hatch. The "horns" on some cases draw oxygen from the water and release waste. Some species attach their cases to seaweed or rocks; others allow them to drift.

range Worldwide.

key sites Cases may wash up on beaches anywhere.

did you know? **many sea creatures** eject their eggs directly into the sea? Other species construct cases that hold the eggs until they are ready to hatch. Cases range from simple sacs to elaborate structures.

The long, spiral, chainlike egg case of a whelk

look for **Sea beans** camouflaged within bits of seaweed and other organic matter along the debris line left after high tide.

A sea bean

Sea Beans

L 1.5" diameter, 0.75" thick **F** **S**

appearance Colors include light brown and reddish brown. Sometimes they have a darker or black band partly around the middle, making them look like hamburgers.

origins These are actually the seeds and fruits of woody, high-climbing tropical trees and vines. They are also called drift seeds. They drop into streams and rivers such as the Amazon and then are carried to the ocean, where they may drift thousands of miles for many years—sometimes currents carry them from the Caribbean and Florida all the way to Europe. Sea beans are very buoyant, due to an internal air pocket. Some types of sea beans are sold as nutritional supplements.

range Along the Atlantic coast and Gulf Coast; Caribbean islands.

key sites Most wash ashore from Miami Beach to Melbourne Beach, Florida.

Driftwood Beachcombers regularly find individual pieces or piles of wood. This is driftwood, which comes from trees or human-made structures that fell or were washed into the water. Unlike plastic, driftwood creates its own ecosystem as it floats in the ocean or sits onshore. A piece of driftwood riddled by both tiny and larger holes has hosted two kinds of worms. Shipworms, which are not really worms but a wood-eating species of clam, bore the larger holes, which are 0.25 inch wide and lined with a calcium material. They are a serious menace, destroying wooden structures such as pilings and wooden boat hulls. Gribbles, which are crustaceans related to shrimp and crabs, make the tiny tunnels. If the wood is in water, these creatures' waste and bacteria produce nutrients that support a small marine community under and around the wood: tiny fish, crabs, and invertebrates that hide from predators. Driftwood washed up on a beach shelters small animals, plants, and birds. It can also trap sand and become the foundation for a new dune.

Sun-bleached driftwood rests on a sandy beach.

ON LOCATION

Calvert Cliffs State Park, Maryland

- Glimpse of the distant past
- Easy-to-find fossils
- Stunning sea cliffs
- Great family activity

Sifting through Chesapeake Bay beach sand at Maryland's Calvert Cliffs State Park can transport you to 12 to 16 million years ago, when the beach was a sea inhabited by whales, dolphins, crocodiles, and sharks. The sea's sharks included the now-extinct, 50-foot-long *Carcharodon megalodon,* whose fossilized seven-inch-long teeth can be found along with fossils of other ancient animals. When the land eventually rose above sea level, creatures fossilized in the sediments were carried up with it. During the series of ice ages that began two million years ago, the Susquehanna River carved a valley to the ocean, when sea levels were 400 feet lower than they are today. As the last ice age started to fade 18,000 years ago, the rising sea filled the river valley and created both the Chesapeake Bay and the cliffs on its western shore.

A Constant Supply of Fossils

The Calvert Cliffs stretch for 30 miles along the bay's western shore, beginning south of Annapolis. Fossils litter the beach because the cliffs are eroding, and they sometimes rapidly dump sand, dirt, and fossils onto the beach. The beach continually washes away, which is why you can still find fossils even though people have been picking them up for years. Archaeologists have found evidence that Native Americans used fossilized sharks' teeth to make scrapers and arrow and spear points at least 8,000 years ago. You can find teeth but no bones because sharks' skeletons are made of cartilage, which does not fossilize well. In addition to holding Miocene epoch sharks' teeth, debris from the

A *megalodon* shark's tooth

Digging for fossils at Calvert Cliffs State Park

cliffs contains fossils of whales, dolphins, turtles, porpoises, and rays. Invertebrate fossils include small crustaceans, clams, oysters, corals, sand dollars, and microscopic algae. Evidence shows that the region's climate during the Miocene epoch was somewhat like that of the Carolinas today.

Maryland's state parks have two major rules for fossil hunting: Never dig for anything in the cliffs, no matter how small the object looks, and never try to climb the cliff, because you could trigger a deadly landslide. The beach is a two-mile walk from the parking lot. You should plan to look for fossils an hour or so before or after low tide. The park's website says that you are permitted to take away fossils, but because fossils help tell the story of the past, "be sure to leave a piece of the story for other visitors." Weekdays are the best time to fossil hunt at Calvert Cliffs during summer since the beach is likely to be crowded on the weekends.

Rounding Out a Fossil Excursion

Before or after fossil hunting at Calvert Cliffs State Park, you could spend at least a couple of hours at the Calvert Marine Museum in Solomons, Maryland, eight miles south of the park. Here, you can see and photograph a life-size diorama of a *megalodon* shark. A time line of the prehistoric epochs, which fills an entire wall and depicts many important fossils from the cliffs, can teach you more about the fossils you have found. You could also visit the Battle Creek Cypress Swamp Sanctuary, 14 miles north of the park. This site is the northernmost U.S. cypress swamp. A quarter-mile boardwalk trail offers a chance to spot some of the swamp's wildlife and to see the 100-foot trees and their "knees" up close. Exhibits in the nature center provide more information on this unique water's-edge environment. Exhibits also include the area's cultural history.

The truth is that we need invertebrates but they don't need us. If human beings were to disappear tomorrow, the world would go on with little change . . . But if invertebrates were to disappear, I doubt that the human species could last more than a few months.

—EDWARD O. WILSON

[2]

INVERTEBRATES

Sea stars and sea anemones are among many
spectacular creatures in tide pools along rocky coasts.

Jellyfish

If you have never seen a jellyfish, you would have a hard time imagining such a creature: a gelatinous blob with long, stringy tentacles. Jellyfish are not fish. They do not have a brain, a central nervous system, eyes, or even a digestive system. They absorb nutrients directly into a digestive cavity. A loose network of nerves triggers poisonous stingers on the tentacles to paralyze prey. The jellyfish's complex life cycle begins with a form called a polyp, which is attached to the sea bottom before it begins to float in the ocean.

A moon jellyfish is identified by the circular structures in its center.

Moon Jellyfish

Aurelia aurita
Diameter up to 18" **S**

appearance These large, circular creatures are nearly transparent, except for a fringe around the edge and horseshoe-shaped gonads in the center, where eggs develop. The "umbrella" has eight parts. Their small tentacles have stinging cells.

food Their diet is mainly small plankton, mollusks, crustaceans, larvae, copepods, rotifers, nematodes, protozoans, diatoms, and tiny eggs.

behavior Moon jellies can swim by pulsating their bell-shaped "umbrella," but this merely enables them to stay near the water's surface. They can travel long distances only by drifting with ocean currents.

range Found near the Atlantic, Pacific, and Indian Ocean coasts, mainly in warm waters—but they are able to live in water temperatures as low as 43°F.

safety tip

Cells on the tentacles of a jellyfish, alive or dead, inject venom when touched. Some types of venom are deadly.

A sea nettle propels itself through the water.

Sea Nettle

genus Chrysaora
Atlantic Sea Nettle diameter up to 6" **S**
Pacific Sea Nettle diameter up to 20" **S**

appearance The Atlantic species is pale, either pink or yellow, and often radiates more deeply colored stripes on its float. The Pacific species has a distinctive golden-brown bell with a reddish tint. The tentacles can extend several feet.

food The diet consists of zooplankton, crustaceans, sea snails, small fish, jellyfish, and even larvae.

behavior A sea nettle's movements are driven by wind, tides, and currents, but it can produce weak swimming motions by contracting and relaxing its body. The tentacles are spread out like a large net with stingers to stun prey.

range Atlantic nettles occur from Cape Cod to the Gulf of Mexico, and in the Caribbean. They are plentiful in Chesapeake Bay in warm weather. Pacific nettles are most common on the California and Oregon coasts.

Portuguese Man-of-War

Physalia physalis
Diameter 12" **S**

appearance This beautiful but venomous "jellyfish" has a large, purple, gas-filled float with a pink line across the top of its crest. Its tentacles can be up to 165 feet long.

food small fish, crustaceans, plankton

behavior The Portuguese man-of-war is not a true jellyfish. It consists of a colony of individual organisms called polyps, which function as a group. The float drifts on the ocean's surface and is carried by winds and currents, which sometimes blow the animal onto the shore. Do not touch a man-of-war on a beach. The tentacles are painfully poisonous.

range Tropical and warm subtropical oceans globally. It is most common in the Pacific and Indian Oceans, where currents often carry it to middle latitudes.

A Portuguese man-of-war stranded on a beach

Sea Anemones & Corals

Sea anemones and corals are animals that look like plants. Both have symbiotic relationships with single-cell green algae, meaning that each organism serves the other. The algae produce food and oxygen for the animals. In turn, the animals keep the algae exposed to the sunlight and protect them from being eaten.

A sea anemone's tentacles await prey as they wave in the current.

Sea Anemones

order Actiniaria
L up to 6' **S**

appearance These animals have column-shaped bodies encircled by tentacles at the top. Species are beautifully colored in fluorescent green, orange, red, or yellow. Tentacles vary greatly in length and shape.

food The diverse diet includes mussels, other small mollusks, copepods, small crustaceans, marine larvae, and marine worms.

behavior Anemones consist of polyps.

Each polyp has a vertical tube with a basal disc attaching it to a substrate, an oral disc at the top serving as a mouth, and stinging tentacles that capture prey. One polyp produces both eggs and sperm; a fertilized egg becomes a larva that develops into a new polyp.

range More than 1,000 species worldwide; mostly in the tropics.

did you know?

clownfish have mucus protecting them from anemones' stings? This allows the clownfish to swim around the anemones.

True Corals

order Scleractinia
L varies widely **S**

appearance Like anemones, individual coral animals are called polyps. Their exterior skeletons have a dazzling variety of shapes and colors that snorkelers and divers love to see. Some species are named for their resemblance to familiar shapes, such as the brain, staghorn, elkhorn, orange-cup, clubbed finger, and lettuce-leaf corals.

food Corals eat phytoplankton and zooplankton, which are tiny plants and animals.

behavior Coral polyps have hard exterior skeletons surrounding the soft animals inside. The polyps can live separately, or individual polyps can divide into clones that connect to each other as colonies. The connected skeletons form reefs.

range Shallow waters in the tropics, deep water in temperate zones.

White tips identify this animal as a fire coral.

Fire Coral

genus Millepora
L up to 24" high **S**

appearance Most fire corals look like trees, other types of leafy plants, or seaweed. Their colors include cream, yellow, brown, red, and purple. Their branchlike and twiglike tips usually have fine white stinging tentacles. If touched, fire coral has an extremely painful and dangerous sting. When dead, the calcified external skeleton can scrape the skin.

food Fire corals eat tiny floating animals and algae.

behavior The soft animals inside the fire coral's skeletons are nearly microscopic in size and are connected by a network of tiny canals. The individual animals are called polyps. For reproduction, they first produce offspring called medusae, which release both eggs and sperm into the water. Fertilized eggs produce free-swimming larvae that eventually attach to a substrate and form new colonies.

range Tropical and subtropical waters.

Chunks of long-dead coral washed ashore

A marine worm works its way into a narrow passage between rocks.

Worms

Calling someone a worm is not considered a compliment, but maybe it should be. This major division of the animal kingdom, commonly known as ringed worms, includes thousands of species in all of Earth's environments. Without these worms, marine, freshwater, and terrestrial food webs would collapse. In food webs, worms eat tiny plants and animals and larger animals eat them.

Marine Worms

class Polychaeta
L up to 9' **F** **S**

appearance Nearly 6,000 species of marine worms are called bristle worms because they have bristles on leglike appendages extending from the body. These cylindrical worms consist of many thin segments. Their heads have eyes along with other sensory organs. Like land worms, marine worms move by alternately contracting and stretching their bodies.

food Some hunt for food, including a few that inject venom into prey; others eat sand or mud to digest detritus in it; still others sift food particles from moving water or use tentacles to pull water with food into burrows.

did you know? **millions of bristle worms** of some species cover the water's surface at night? In some places they spawn en masse annually. A certain water temperature during the new moon seems to trigger this spawning.

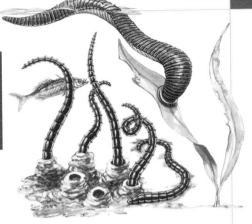

safety
tip

Leech bites bleed more than other wounds do because leeches secrete an anticlotting enzyme into the bloodstream. Always treat leech wounds to avoid infection.

behavior Separate sexes send eggs and sperm into the water, where fertilization occurs. When larvae hatch, they have mouths and digestive tracts. Gradually they add segments and grow more and more elongated. Larvae are an important food source for fish, wading birds, and other predators. Adults live in crevices and on sea plants.

range All oceans, in bottom mud, tidal flats, and even in extremely high temperatures around hydrothermal vents at the ocean bottom.

Freshwater leeches in a pond

Freshwater Leeches

class Hirudinea
L 0.2–10" **F**

appearance Leeches are segmented worms with suction cups at each end. Their bodies are flattened and much wider than they are thick, and they are usually dark in color, often brown. Some species have no markings, while others have spots and stripes. They may be well camouflaged.

food Leeches eat the blood of fish, frogs, turtles, and mammals; sometimes they dine on amphibian eggs and snails.

behavior A leech's digestive tract has lateral pouches that hold more than the animal's own weight in blood. The leech can digest so much blood that the body expands up to five times its normal size. Some species can go a year without eating. Leeches detect their prey through vibrations. An individual is both male and female; it lays eggs inside small cocoons in the muddy bottom.

range Fresh water, including rivers, lakes, and swamps.

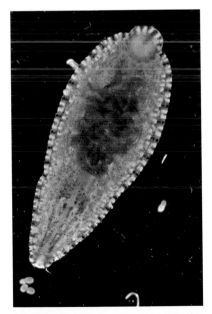

A turtle leech carries its young on its underside.

Clams & Oysters

Clams and oysters have several things in common: People like to eat them; they are bivalves, meaning that they have two shells held together by a hinge; and they require water. The word *clam* is sometimes used as a name for an edible bivalve, but in North America *clam* usually refers to all bivalves that burrow in sand or mud. Bivalves that attach themselves to a substrate are known as oysters. Unlike clamshells, oyster shells are lumpy and irregular in shape.

A razor clam shell with both sides still attached

Clams

family Veneridae
L 5-6" F S

appearance Clams have thick, two-part shells, some with ridges or patterns, some thick, some thin.

food These mollusks filter out microscopic food from water.

behavior A clam is usually found buried not far beneath the surface. It extends its syphon into the water around high tide. A gill structure filters out food from the water; the syphon also expels waste. The spawning season depends on water temperature. A group of clams sheds eggs and sperm into the water at the same time. After fertilization, typical life stages are two free-swimming larval forms that last for varying amounts of time before settling in to their preferred habitat as adults.

range Various species in all seas, including the Arctic Ocean and the Southern Ocean around Antarctica.

safety tip

Oysters from unpolluted but warm seawater can harbor the *Vibrio vulnificus* bacterium, which can cause severe illness. Thoroughly frying, stewing, or roasting the oysters will kill the bacterium.

Bounty of the sea: clams, oysters, and fish

Ready for an oyster meal

Oysters

family Ostreidae
L up to 8" F S

appearance Oyster shells are often distorted from being cemented to other objects. Barnacles and algae often attach to them, which makes it hard to distinguish the oyster shell from other organic material.

food Oysters eat microscopic organisms filtered from water.

behavior Oysters pump water through their body cavities and cause rhythmic waving of cilia (hairs on their gills), which take up oxygen and filter food. A muscle opens and closes the shell. An ability to separate food from silt allows oysters to survive in turbid water and to remove some pollutants. They help keep water clean. The animal reproduces by releasing sperm and eggs into the water, where fertilization occurs. Free-swimming larvae may drift as far as 800 miles, attaching permanently to a hard surface.

range Intertidal areas of all of the Earth's oceans.

Gathering Wild Shellfish

Hunting wild shellfish is good way to enjoy a day at the water's edge. Before starting, be sure to check state regulations regarding when and where you can hunt shellfish. In addition to learning where to go, you need to find out whether any areas are temporarily off-limits because local shellfish might contain toxins. Ordinary pollution is a serious hazard, but nature also produces harmful toxins. In some places shellfish may consume algae that carry dangerous poisons. Cooking does not destroy these toxins the way it kills bacteria. There are three types of shellfish poisoning: neurotoxic, paralytic, and amnesic. The last two types can be fatal. If you feel ill after eating oysters, seek medical attention immediately. Always keep your shellfish cool as you hunt; a netted bag submerged in the water inside a floating inner tube works well. You will also need ice to keep your meal from spoiling—unless you are lucky enough to cook and eat by the water.

The end of an enjoyable day hunting shellfish

Scallops & Mussels

All species of scallops have similar shells and life cycles, and they may live in salt water or brackish water. The main difference is that bay scallops, which live in estuaries and salty or brackish bays, are smaller than sea scallops, their ocean cousins. In contrast, mussels are not so clear-cut. In North America the word *mussel* is generally used for several very different families of saltwater and freshwater bivalves that attach themselves to a substrate by threads. Zebra mussels are not related to these species.

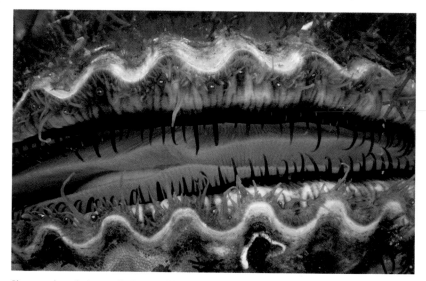

Close-up view of a bay scallop's rows of blue eyes

Bay Scallops

Argopecten irradians
L up to 1.5" **F** **S**

appearance Colors of the upper shell range from dark gray to brown, and some have yellow, orange, or red highlights. Many ridges radiate like a fan from the base.

food Bay scallops filter microscopic nutrients from the water.

behavior Bay scallops lie on the bottom in shallow waters, usually where eelgrass is present. Their eyes can detect nearby movement, warning of the presence of predators. Scallops mature and spawn at about one year old and may live as long as 20 months. Scallops move by using a specialized muscle, which rapidly

Bay scallops come in different colors.

A couple sorts through mussels.

opens and closes the upper and lower shell, ejecting a pulse of water.

range High-salinity bays and estuaries along the Atlantic coast and Gulf Coast, through the Caribbean region.

Marine Mussels

family Mytilidae
L up to 6" **S**

appearance Mussel shells are generally pear shaped and are often dark blue, blackish, or brown, with a silvery interior.

food Mussels eat microscopic nutrients filtered from water.

behavior Mussels attach themselves to a hard substrate by a mass of threads called a byssus, which the animals secrete. The mussel also uses these threads to entrap, immobilize, and starve predatory mollusks such as dog whelks. Usually, mussels clump together in colonies on rocks or pilings of piers and marinas. Fertilization occurs outside the body. Larvae drift for three weeks to six months before attaching to a hard surface.

range Abundant in the intertidal zone in temperate seas globally. Some species live in salt marshes, and others thrive in pounding surf.

Invasive Zebra Mussels

Zebra mussels invaded Europe from Russia in the 18th and 19th centuries and reached North America's Great Lakes in the 1980s. Most of the invaders arrived in ballast water carried by transatlantic ships. These mussels ruin ecosystems by crowding out native mollusks. Thick colonies disrupt infrastructures when attached to docks, boats, and insides of intake pipes to municipal water plants and power plants. Since the 1980s, zebra mussels have spread into large areas of North America, including southwestern lakes, and they are expected to continue spreading because no safe method of killing them is known.

Zebra mussel shells

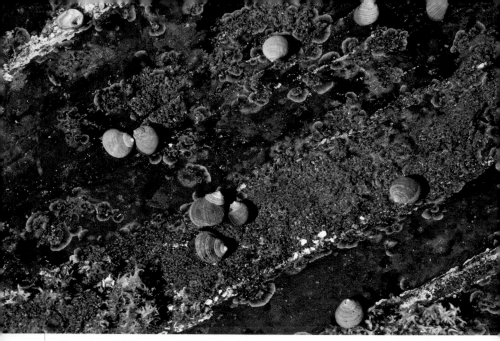

Common periwinkles generally live in colonies on rocky shores.

Mollusks

The 85,000 living species of mollusks make up a tremendous variety of saltwater and freshwater creatures that have shells, such as snails and clams. But mollusks also include cephalopods—octopus, squid, and cuttlefish—which have a head, a container for internal organs, various types of tentacles, and a muscular foot. The cephalopods are believed to have taken an evolutionary path away from mollusks with exterior shells by trading their heavy shells in exchange for increased mobility.

Periwinkles

Littorina littorea
L up to 1.5" **S**

appearance These small shells are sharply pointed except when eroded by surf. Six or seven spiral whorls have fine ridges and wrinkles. Colors vary from grayish to gray-brown, often with dark spiral bands. The inside of the shell is chocolate brown.

food A periwinkle's diet consists of mostly algae, along with invertebrates such as barnacle larvae.

behavior Reproduction is annual, with internal fertilization. Females release up to 100,000 eggs into the sea. Larvae drift as plankton for four to seven weeks before settling to the bottom as adults in intertidal zones on rocky shores, in tide pools, or in muddy habitats such as estuaries. The edible periwinkle has been an important food source for Europeans for thousands of years.

range The common periwinkle is the most abundant mollusk along the Atlantic coast, from New Jersey as far north as Labrador.

did you know ?

periwinkles are invasive? After arriving from Europe in the mid-19th century, they have changed North Atlantic ecosystems by altering distribution and abundance of algae and displacing native species.

Abalones

genus Haliotis
L 4–8" **S**

appearance The shell may look like that of a clam, but an abalone is a type of snail with an oval, ear-shaped shell pierced by large respiratory holes. Collectors prize the shell because of its interior's bright iridescence.

food The diet is mainly red or brown algae and kelp.

behavior Abalones cling tightly to underwater rocks with muscular feet. Females lay millions of eggs at a time. The eggs hatch as microscopic, free-swimming larvae that drift with currents for about a week before settling to the bottom. Many people consider abalone meat a great delicacy.

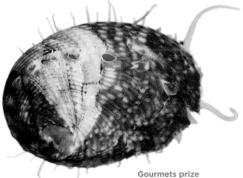

Gourmets prize red abalone.

range Most species are found in cold waters along shores of western North America, Japan, New Zealand, South Africa, and Australia.

Red Octopus

Octopus rubescens
L mantle up to 40", arms up to 16" **S**

appearance Like all octopus species, this one—although it is called red—can change its color and texture to match its surroundings. Colors can vary from a deep brick red to brown, white, or mottled mixtures.

A red octopus crawls over a reef.

food Octopuses eat lobsters, crabs, mussels, snails, and fish.

behavior An octopus uses long tentacles to grasp prey and pry apart shells. Females guard their egg clusters in shallow water, including tide pools. The breeding peak is in August and September; young hatch in six to eight weeks, spend a brief period as plankton, and then settle in kelp beds as juveniles.

range From Alaska along the Pacific coast to Baja California.

Freshwater Mollusks

North America's streams, rivers, and lakes are homes to approximately 300 species of freshwater mussels, while all of Europe has only 12 species. North America also has 650 species of freshwater snails, second only to Southeast Asia. The continent's freshwater environments are among the world's richest, and these mollusks are important parts of the food webs of birds, fish, and other animals. They are threatened by habitat degradation arising from dams, channelization and dredging of rivers, sediments washing into the water, and pollution from mine runoff, industrial wastes, and chemical spills.

Virginia river snail

Virginia River Snail

Elimia virginica
L 1.5" **F**

appearance This snail's shell is elongated, with a high, narrow spire. Shells are variable in coloration, although they are generally yellow to chestnut. Some, especially juveniles, have two darker brown spiral bands.

food The diet includes a mixture of algae, cyanobacteria, microbes, and detritus attached to submerged surfaces.

behavior The preferred habitat is in freshwater rivers and streams that have gravel or cobble bottoms—thus, siltation is a threat. In spring and summer, the female deposits eggs on hard substrates. This species grows more slowly than many other types of freshwater snails.

did you know ? **snails are important** to freshwater environments? Other animals, including fish, reptiles, and birds, eat them. After eating algae and detritus, snails return nutrients to the environment in their waste.

An apple snail munches on water lettuce.

range native to the United States in rivers east of the Appalachians, from North Carolina to Connecticut. This snail spread to the Lake Ontario Basin in the 19th century via the Erie Canal.

Freshwater mussels often form aggregations.

Freshwater Mussels

order Unionoida
L up to 12"

appearance There are approximately 300 North American species, whose colors range from yellow to green to brown to black, sometimes with dark rays extending from the base.

They come in many sizes and shapes: round, irregular, triangular, and squarish. Some are smooth, some have fine grooves, and others have rough ridges.

food The diet includes plankton and microscopic organisms filtered from water.

behavior The male releases sperm into water, and they are drawn into the female as she filters water for food. Eggs develop in gill pouches. Larvae attach themselves to a host as a parasite until they drop off to begin an independent life.

range permanent lakes, rivers, canals, and streams with a constant source of cool, clean water. Many species are endangered or threatened.

did you know ? **health officials advise** you not to eat freshwater mussels? They filter contaminants that accumulate in their tissues and are toxic to humans

A female southern rainbow mussel flares its gills to tempt a host fish.

ON LOCATION

Bay of Fundy, New Brunswick, Canada

- World's highest tidal ranges
- Walk on the ocean's bottom
- Millions of birds
- Whale-watching

Most of the time when you go to the beach, you don't have to worry about the tide coming in unless it's a beach where you can park your car below the high-tide line. Beaches around the Bay of Fundy are an exception. This roughly 180-mile-long bay, which runs southwest-northeast between northeastern Maine and Canada's New Brunswick Province to the north and Nova Scotia Province to the south, has the world's largest range between high and low tides. Water squeezing through the bay's 50-mile-wide opening makes tides higher in the two arms that form a Y at the upper end than they are at the mouth. The highest tides occur at Burntcoat Head, Nova Scotia, near the end of the Minas Basin, the bay's southern arm. The average difference between high and low tide there is 48.5 feet. The world record tidal range was set at Minas Basin on February 8, 1997, when there was a range of 54 feet, 7 inches. On that day, the positions of Earth, the sun, and the moon were most favorable for a wide range.

Uncovering the Bay's Floor

At low tide water drains from more than 620 square miles of the bay's floor and uncovers millions of plants and invertebrates living in the mud and sand. They are adapted to living half of each day underwater and half of the day exposed to the air. Vertebrates such as fish and marine mammals follow outgoing tides into deeper water. The remaining plants and animals attract millions of land

The Bay of Fundy is known for its tidal range.

An American lobster nests in seaweed in the Bay of Fundy.

and shorebirds to feed—and hundreds of bird-watchers to observe them. Each year's big bird event is the mid-July arrival of thousands of semipalmated sandpipers making their only stop during migration from the Arctic to Central and South America. Other birds stopping around the Bay of Fundy include dabbling ducks and geese, which forage in the marshes and other wetlands during their spring and fall migrations. Black ducks, loons, mergansers, common eiders, white-winged scoters, old-squaws (a type of sea duck) winter around the bay.

Up Close With Bay Wildlife

You can walk on the exposed bay bottom at low tide, but do not disturb feeding birds. Before going out you need to find out when the tide will begin rising—and it might rise faster than you can run. To experience the tides, you could spend the 6 hours, 13 minutes between high and low tides at the same place. To learn when tides are high and low at a place you plan to visit, search the Internet for "Bay of Fundy tide tables." You'll see the biggest tidal ranges around the times of the new and full moons. When an incoming tide meets a river's normal outgoing flow, it forms a tidal bore, which is a wave travelling up the river. These are especially strong on the Shubenacadie River near South Maitland, Nova Scotia, and the St. John River in the middle of St. John, New Brunswick. Companies in both places offer boat rides through the waves that the bores create as they go over rapids and sandbars.

To enjoy the bay's biggest wildlife, you can take whale-watching cruises from ports on both sides of the bay near its mouth. Large whales seldom swim far up into the bay.

Crabs

Crab meat is a staple on menus located hundreds of miles from the salt water where the animals live. Crabs are entertaining to watch as they scramble sideways in the wild. Hermit crabs that live in seashells are commonly kept as pets. Learning about crabs and their environments is a good introduction to bay and estuary food webs and ecosystems.

Blue crab

A blue crab is white on the underside.

Females migrate to high-salinity water to hatch eggs. After mating, the female extrudes fertilized eggs into a cohesive mass, or sponge, attached to her abdomen until larvae emerge. Adults "molt" their shells.

range Atlantic coast from Nova Scotia to Argentina; Gulf of Mexico, Caribbean Sea. Rare north of Cape Cod except after consecutive warm years.

Rock Crab

Romaleon antennarium
L up to 9" **S**

appearance This species has a fan-shaped deep red or brown shell, which can vary to shades of orange or gray. The underside of the crab is spotted. The claws are black tipped, and the front claws are very thick. Walking legs are usually hairy.

Blue Crab

Callinectes sapidus
L up to 9" **F S**

appearance The top of the shell is dark green, the underside is white, and the front claws are blue. The crab has six legs; the front pair is for grasping, and the rear pair is paddle shaped. Eyestalks protrude from front.

food This crab eats bivalves, small fish, worms, plants, carrion, detritus, and other blue crabs.

behavior The blue crab lives in low-salinity estuaries, bays, and tidal rivers.

Pacific coast rock crab

Crabbing is a livelihood for many people.

Crabbing is a recreational activity that does not require expensive equipment. First, learn whether the state you are in requires a license or places restrictions on when, where, or how you may try to catch crabs. The rules might have changed since you were a child. Using a crab trap is the usual method. A popular way, especially with children, is to tie a chicken neck to a long cord and lower it into the water from a pier or a boat. When you feel a nibble, slowly pull up the cord and try to catch the crab in a long-handled net before it drops off the bait.

food The rock crab eats a variety of bivalves, snails, sea stars, and other crustaceans, including hermit crabs. It reaches a hermit crab by gradually chipping away the seashell in which the hermit lives.

behavior Rock crabs are common in the low rocky intertidal zone, often under rocks. They are typically active by night. Mating takes place in spring or fall. Egg-carrying females are most often seen during winter. Eggs are extruded 11 weeks after mating. After hatching, the larvae go through six stages as plankton before they reach the adult stage.

range U.S., Canada west coasts.

Ghost Crab

Ocypode quadrata
L 9" wide **S**

appearance This small species is straw colored or grayish white, with a rectangular shell. It has hairy legs, which are often difficult to see because they are sand colored and translucent.

food This crab is a scavenger of organic matter. It feeds on mollusks, insects, plants, detritus, and other crabs.

A ghost crab's eyestalks are otherworldly.

behavior It is a largely nocturnal species, avoiding predation by gulls when it is feeding. It is rarely seen during daylight, when burrowed in soft sand from near the high-tide line to a quarter-mile away. Females lay eggs in the water, and larvae drift for four to six weeks before returning as young crabs.

range Sandy beaches along the Atlantic Ocean, Gulf of Mexico, and the Caribbean. Overall range from Rhode Island to Brazil.

look for **ghost crabs** after dark. If you point a flashlight at it, the crab will stand still. During the day, look for their golf ball–size burrow entrances.

A hermit crab will occupy any kind of shell that fits.

Hairy Hermit Crab

Pagurus hirsutiusculus
L up to 2.8" **S**

appearance This species ranges in color from olive green to brown to black. Its body has many projections that look like hairs. The legs often have white and blue bands, and the antennae are grayish brown with white bands.

food The hairy hermit is a scavenger of detritus, including dead animal and plant material.

behavior Like almost all 1,100 hermit crab species around the world, this one carries an abandoned, lightweight shell. If frightened, it may abandon the shell. A hermit crab needs to select increasingly larger shells as it grows. The female broods her egg clutches for two to three weeks; she holds them on her abdomen with modified legs. After hatching, the larvae go through four stages before they mature and settle in to adult life in a snail shell on the seafloor.

range Pacific coast from Alaska to southern California. The hairy hermit is the most common hermit crab around San Francisco Bay.

> **look for** **hermit crab** activity at night. These crabs are nocturnal scavengers. They spend their daytime hours hidden under a rock or in a dark crevice.

Fiddler Crab

genus Uca
L up to 2" **F** **S**

appearance One of the male's claws is enormously oversize; the female's claws are about equal in size. Most species are drab tan or brown, but some have brightly colored legs and claws. Movements of the male's small

did
you
know
? **fiddler crabs** shed their shells as they grow? A new leg claw will replace a lost one when they molt. If a male loses his large fiddle claw, he will grow one on the opposite side after his next molt.

A male fiddler crab's large claw is conspicuous.

claw against its large claw resemble someone playing a fiddle.

food These crabs salvage anything edible from sediment, including algae, microbes, fungus, and organic detritus. They feed by picking up small amounts of sediment with their claws and using mouthparts to scrape off food. Then they put the inedible sediment back onto the ground as a deposit called a feeding pellet.

behavior Habitats include sandy areas, intertidal mudflats, lagoons, swamps, brackish marshes, and mangroves. These crabs live in colonies and are active during the day.

Females carry up to thousands of tiny eggs under their abdomens. Larvae live in open water and pass through several molt stages as they grow.

range Various species are found along the Atlantic and Pacific coasts, as well as in oceans worldwide.

Hermit Crabs as Pets

Most people get their first—and probably only—hermit crab from a beachside souvenir shop. If these crabs are not cared for correctly, they may not live long. However, some lucky hermit crabs end up with enthusiasts who keep them in a "crabarium" or a "crabatat" that is kept clean and at the proper temperature. Hermit crabs that are given proper care can live for many years. Keeping a hermit crab healthy is not easy. Most hermit crabs that are available for purchase are tropical. Proper care begins with keeping them in a warm and humid environment, but they require additional attention—especially a proper diet. Hermit crabs are definitely not hermits. They are extremely social animals, and they will have a more natural environment if you keep several together rather than having just one.

A hairy hermit crab partly inside a shell

A crayfish scavenges for food on a muddy bottom.

Other Crustaceans

Crustaceans are a large group of animals including shrimp, crabs, lobsters, crayfish, and barnacles. Most of them live in salt water, but some are at home in freshwater or on land. All crustaceans have antennae on the head and "arms," and they use these body parts for feeding. An exterior skeleton covers the two main segments of the body, which have various types of appendages used for walking, swimming, and carrying eggs. It may come as a surprise that crustaceans are closely related to insects and spiders.

Red Swamp Crayfish

Procambarus clarkii
L up to 5" **F**

appearance This species is dark red, with raised bright red spots covering its body and claws and a black, wedge-shaped stripe on the top of the abdomen.

food Crayfish are omnivorous. They eat living and decomposing plant matter, seeds, algae, invertebrates, and sometimes small fish.

behavior Preferred habitats are marshes, swamps, ponds, slow-moving rivers, and streams. The animal mates

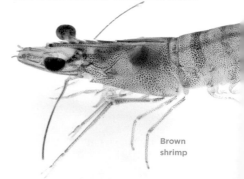

Brown shrimp

in autumn and lays its eggs in spring and early summer. Hatchlings remain attached to the female for several weeks and through two molts. Crayfish can survive long dry spells by remaining in a burrow or by crawling overland to seek water.

range Gulf Coast from northern Mexico to Florida Panhandle, north to Illinois and Ohio.

Brown Shrimp

Farfantepenaeus aztecus
L up to 9" **S**

appearance This shrimp is brownish, with ten slender, complex legs. Adults have long, thin antennae; a shell-like structure called a carapace covering its internal organs; and a long, segmented shell extending to a finned tail.

food The diet consists of various crustaceans, worms, mollusks, larvae, and organic debris.

behavior Brown shrimp mate in deep water. The females lay 50,000 to one

million eggs; larvae go through several stages in two weeks to become small adults. Young mature in estuaries, move to ocean.

range Worldwide in tropical and temperate waters. In U.S. waters, from Cape Cod to the Gulf of Mexico and the Caribbean.

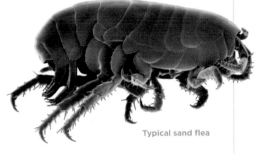

Typical sand flea

Sand Flea

genus Emerita
L up to 1.5" **S**

appearance Sand fleas are much too large to be true fleas and are often called mole crabs (though they are not crabs, either). They have a brownish or grayish, barrel-shaped body with short, leglike appendages and feathery antennae. They are often seen rolling in the swash, pushed ashore, and then burrowing into the sand. Watch closely, because they can bury themselves completely in as little as 1.5 seconds.

food The diet consists of microorganisms filtered from water.

behavior Males are typically smaller than females. In some species, a male lives attached to the legs of the female. The larvae float as plankton for more than four months and are carried long distances by ocean currents.

range U.S. Atlantic, Gulf, and Pacific coasts; Atlantic coast of Africa.

Barnacles

Barnacles may look like mollusks with shells over their bodies and no legs or other appendages. Nevertheless, like all crustaceans, barnacles have separate heads and bodies with appendages for capturing food. They come in a tremendous variety of sizes and shapes and can be found anchored to rocks, boats, whales, seashells, and wharf pilings. Unlike other crustaceans, a barnacle can withdraw its appendages into its shell to avoid drying out when it is not immersed in the water during low tide. Barnacles feed when immersed.

Common barnacles underwater, with shells open at top

Common Barnacle

Semibalanus balanoides
L up to 0.6" **S**

appearance This tiny species is white or grayish, sometimes with shades of purple or pink. Its shape resembles a cone with the top cut off.

food This barnacle filters water for microscopic creatures.

behavior Embryos develop into a first larval stage inside their parents' bodies and then are released as drifting plankton. After six molts, they become forms that search for a suitable substrate for attachment. When they find one, they release a glue that holds them to the substrate, and then they undergo metamorphosis

safety tip **Barnacles have sharp edges** that can cut and scrape you. Be cautious around rocks with barnacles attached or when swimming near pilings.

different species of barnacles on rocks and pilings, from the splash zone above the high-tide level down to fairly deep underwater.

Barnacles uncovered by water, with shells closed

The feathery appendages face toward land and comb the water for plankton carried by waves returning to the sea. The animals require moving water to push food toward them. Larvae drift with ocean currents.

into their adult form. They remain attached for the rest of their lives.

range Intertidal zone in the world's northern oceans. In North America, from Labrador to Cape Hatteras on Atlantic coast, Alaska to British Columbia on the Pacific coast. Average monthly ocean temperature must be above 45°F for breeding.

range There are two species. One is found on the Pacific coast of North America, from Alaska to Baja California. The other lives on the European coast from Spain and Portugal to the United Kingdom.

Gooseneck barnacles filter food from the water.

Gooseneck Barnacle

Pollicipes species
neck up to 4" long, shell up to 2" wide **S**

appearance Also called goose barnacles, these animals look something like bivalve seashells (but they are not related to shells). A long "neck" stalk attached to a rock holds the white body firmly to the substrate. Feathery, comblike appendages extend from the shell.

food Diet consists of microscopic organisms filtered from water.

behavior Colonies live in crevices on rocky shores with strong waves. The animals are anchored by a tough, flexible stalk that contains the ovaries.

Sea Stars & Cucumbers

Many people say that sea stars are cute, but they call sea cucumbers creepy after learning that they can expel some of their inner organs to entangle would-be predators (the organs grow back). Both groups of species illustrate how animals have evolved over millions of years to fit in to particular environments. Several species of sea stars and cucumbers can be found in tide pools.

Sea Stars

class Asteroidea
Diameter up to 10" **S**

appearance An estimated 2,000 species of sea stars (commonly called starfish, although they are not fish) live around the world. Various numbers of arms, most often in multiples of five and as many as 50, are connected to the central part of the body. Colors and patterns vary as well, from drab brown to bright red, pink, orange, yellow, green, lavender, and blue.

food Sea stars feed on slow-moving or anchored prey such as barnacles, clams, mussels, snails, sea urchins, and other sea stars.

behavior Sea stars live in tide pools on rocky shores and on deep ocean bottoms. Their bony, calcified skin protects them from most predators, and they are famous for a capability to regenerate severed or damaged arms. A primitive eye at the end of each arm allows the animal to sense light and dark and helps in detecting prey or predators.

range From the intertidal zone to deep water in all seas, including the Arctic Ocean and the Southern Ocean around Antarctica.

did you know?
most sea stars that people see on beaches have five arms in a typical star shape? But other species with more arms, such as brittle stars and basket stars, can live thousands of feet underwater.

A sea star with ten arms

A typical five-pointed sea star

look for sea cucumbers in tide pools and low-tide zones among rocks, where waves and tides bring a regular supply of their food.

Sea Cucumbers

class Holothuroidea
L up to 6.5' S

appearance These creatures are elongated and flexible. They resemble large caterpillars or soft cucumbers. They are often black, brown, or olive green, but some are bright red, orange, or purple. Their mouths are surrounded by a ring of 8 to 30 short tentacles. They propel themselves across the seafloor on tiny tubular feet.

food The diet includes tiny particles such as algae, minute aquatic animals, and organic waste materials.

behavior Sea cucumbers are very sluggish. They swallow large amounts of sand, filter food from it, and then recycle their waste into the ocean as earthworms do on land. When threatened, they can discharge sticky threads to deter predators. They can also regenerate missing parts of their bodies. During reproduction, eggs and sperm are released into the water. Young go through larval states before reaching adulthood.

range Seafloors worldwide, from shallow coastal waters to depths as great as five miles.

A sea cucumber forages in a tide pool.

Pet for an Hour: Sea Star You should not plan to pick up a pretty sea star from a tide pool and take it home to admire in a bucket of seawater. First, taking the sea star from its environment might be against the law. Second, unless you have already set up a seawater aquarium, the animal will quickly die. Serious sea star hobbyists maintain seawater or even coral reef aquariums. This can be a fascinating experience and a great way to observe the day-to-day lives of sea creatures. You cannot simply fill a fish tank with tap water and sprinkle table salt in it; this would kill any sea creature. An Internet search could get you started on a new hobby of keeping a sea aquarium.

ON LOCATION

Fitzgerald Marine Reserve, California

- Tide pools
- Weird invertebrates
- Helpful guides
- No admission charge

If you have never visited or explored North America's Pacific coast habitats, Fitzgerald Marine Reserve is the perfect introduction. Here you'll find thriving life in tide pools, mixed tides, cold ocean water, and foggy summer days. Tide pools are rocky ocean-beach basins filled at high tides by water that remains in the pool when the tide goes out. This water keeps marine plants and animals from drying out and dying until high tide brings a fresh supply of water, food, and oxygen. The only other way to see many of the life-forms you'll find in a tide pool is to dive to an ocean bottom.

Life in Tide Pools

Invertebrates in tide pools include sea urchins, sea stars, sea anemones, mussels, barnacles, worms, and

Low tide at Fitzgerald Marine Reserve in California reveals marine life that thrives in tide pools.

limpets. These habitats show some of the many ways in which ocean life has evolved to live in harsh environments. Geology, global ocean currents, and weather patterns account for the West Coast's rich ocean life and unique habitats.

The Pacific coast, where mountains come down to the sea and the ocean bottom drops off rapidly, offers no gently sloping shelf for wide sand beaches like those in the East. The offshore California Current and upwelling water keep ocean water temperatures close to 55°F, even in the middle of summer. In comparison, summer water temperatures are close to 80°F off the Chincoteague National Wildlife Refuge at roughly the same latitude in Virginia. Prevailing winds push water away from the Pacific coast, and this causes cold, nutrient-rich water to rise to the surface. Inland, over California's Central Valley, solar heating causes air to rise. Air from over the ocean flows inland to replace the rising air. As it blows over the cold ocean it cools, and some of its humidity condenses into the fog and clouds that move over coastal areas (this phenomenon is often called the June gloom). Foggy summer days with low clouds are rare on most of the Atlantic coast. On the West Coast, sun often evaporates the fog and clouds in the afternoon, but by then their shade has protected exposed tide-pool water, plants, and animals from the sun during low tides.

A giant green sea anemone with other invertebrates

Visiting the Reserve

The term *mixed tides* refers to the Pacific coast pattern of a lower (or negative) low, a high, a low, and a higher high-tide pattern during each 24-hour, 50-minute, tidal day. Since the period around a lower low tide is best for exploring tide pools, you should plan a visit around this time. The Fitzgerald Marine Reserve in Moss Beach is roughly a 40-minute drive south of San Francisco on California's Pacific Coast Highway. The State of California owns the reserve, which San Mateo County manages as a county park and nature preserve. In addition to tide pools on three miles of beaches, it offers marshes, bluffs that are gradually eroding, cliff-top cypress and eucalyptus forests, picnic tables, restrooms, a parking area, interpretative exhibits and programs, and volunteer naturalist guides who will answer your questions and offer suggestions on what to look for. While you can look as much as you want, you cannot take any of the more than 200 species of animals and 150 species of plants found in the reserve. In fact, you should not pick up any animals or move any rocks, which could crush animals. Female harbor seals sometimes give birth at the reserve from February through May, a reminder that not all of the reserve's wildlife is in tide pools. If you see a lone seal pup, stay away. It's probably waiting for its mother to return from foraging. Mothers avoid pups if humans are near them.

One evening, when we were about ten miles from the Bay of San Blas, vast numbers of butterflies, in bands or flocks of countless myriads, extended as far as the eye could range . . . The seamen cried out "it was snowing butterflies."

—CHARLES DARWIN, *THE VOYAGE OF THE BEAGLE*

[3]

INSECTS
&
OTHER
INVERTEBRATES

An Acadian hairstreak butterfly feeds on the nectar of milkweed blossoms.

Dragonflies

Except for their size, dragonflies look much as they did 300 million years ago, when they were larger than dinner plates. Their independently moving wings let them hover, reverse direction, and chase prey at speeds up to 35 miles an hour. Very large compound eyes provide almost a 360-degree range of vision. After hatching from eggs, these insects may spend years as naiads—aquatic larvae—but may live for mere weeks as flying adults.

Common Whitetail

Plathemis lydia
L 1.6–1.9" F

appearance The male has a thick, white abdomen and a wide dark bar on its wings with a smaller spot near the body. The female's abdomen is thinner and brown with yellow spots, and she has three dark marks on each wing. The naiad is dark brown and just under an inch long.

food The species feeds on soft-bodied flying insects, including mosquitoes, flies, butterflies, mayflies, and termites.

behavior The male often rests with its head lowered, its abdomen higher, and its wings drooping forward. At water margins the female dips in the tip of her abdomen to deposit eggs.

A common whitetail on a pickerelweed

range Throughout eastern and central United States north to Nova Scotia and on West Coast from California to British Columbia. Frequents still and slow-moving water.

Common green darner

Common Green Darner

Anax junius
L 2.8–3.1" F

appearance The male has a brownish green to yellowish green thorax, and slender abdomen that is bright blue changing to green. The female is yellowish green with dark brown on the abdomen. The wings have yellow edges. The green darner has a very large, greenish naiad up to 1.9 inches in length.

food This dragonfly eats all manner of flying soft-bodied insects, including mosquitoes, flies, midges, butterflies, moths, caddis flies, and stone flies. The naiad can capture tadpoles and small fish.

food As the name suggests, this species preys on other dragonflies—including those of similar size, like the common green darner—as well as damselflies and larger butterflies, such as swallowtails and monarchs.

behavior This species is a fierce hunter. It often waits in ambush for prey by hovering overhead. The last segments of the male's tail curl under its body in flight and make a J shape. It uses its legs to rotate monarch butterflies to avoid the most toxic parts.

range Found around moving waters and also away from water in eastern North America.

Dragonhunter

behavior The male is highly territorial, patrolling and defending areas on shores of ponds and lakes. The female deposits eggs in slits cut into stems of aquatic plants.

range One of the most abundant species, found throughout United States and southern Canada. Migrates south in fall, and offspring migrate back north in spring.

look for	**a blue, green, and black target** on the forehead of both male and female common green darners, as well as huge eyes that span each side of the head and meet in the middle.

Dragonhunter

Hagenius brevistylus
L to 3.5" **F**

appearance Features, which are similar in both sexes, include a yellow face with widely spaced green eyes and a black thorax with diagonal yellow stripes. Because the dragonhunter has a slightly clubbed tail, it is also called the black clubtail. The naiad is unusually wide and flat.

Dragonflies mating on the wing

Dragonfly Mating

Mating time finds many dragonflies on the wing, as usual. The varied species have complex mating rituals and equally complex reproductive parts. Among green darners, a receptive female allows the male to clasp her behind the head as she reaches for the sperm packet that he has transferred to the front of his abdomen, and their bodies bend inward to form a wheel. They may stay attached while the female deposits her eggs in water, or the male may stay close to ensure that no other male can mate with her.

Damselflies

Damselflies and dragonflies have similar body types, life cycles, lifestyles, and ancient lineages. Damselflies' smaller size and daintier features belie a similar ferocity. Sharp, cutting mouthparts and strong, closely placed legs allow them to catch, grasp, and make short work of prey. Naiads tend to be small and slender.

Ebony Jewelwing

Calopteryx maculata
L 2.25" **F**

appearance The male has a striking color combination—a black head, a metallic green body with a blue cast, and black wings. The female has a dark green body and smoky brown wings with white spots near the tips.

An ebony jewelwing on a leaf

food The adult eats various small insects, particularly aphids and gnats. The naiad eats small aquatic insects.

behavior This species sits horizontally on leaf or twig perches along the shore. It may spend hours or days on the same perch. The male displays to the female by raising his forewings and abdomen to show a pale underside. The male guards the egg-laying female, but both sexes take multiple mates.

range Found along forest streams and large rivers with moderate currents. Occurs in eastern and central U.S. and in southeastern Canada.

Northern Bluet

Enallagma annexum
L 1-2" **F**

appearance The male is predominantly blue and has black rings on the middle abdomen. The female's color varies from bluish to greenish yellow to tan to brown. The wings are clear and thin, and they appear stalked. The naiad is about one inch long.

food Adults eat a wide variety of soft-bodied insects, including mosquitoes, mayflies, flies, small moths, and aphids. The naiad eats mosquito and mayfly larvae and other aquatic insects.

behavior This species commonly perches on vegetation. The male takes

Common blue damselfly, with its compound eyes

the female away from the breeding site to mate. The female deposits eggs on surface and underwater vegetation.

range Found in ponds, lakes, marshes, and streams from Alaska across Canada, south to Virginia, west to California. Largely absent from southeastern U.S.

American Rubyspot

Hetaerina americana
L 1.5–1.75" F

appearance The thorax is bright metallic red or reddish bronze in the male and metallic green or bronze in the female. Males have a ruby spot at the base of clear wings. A female's spot may be more brown than red. The male is slightly bigger than the female, but the female's abdomen is thicker. Naiads are about one inch long.

food Adults feed on small, soft-bodied insects such as mayflies. Naiads eat mosquito larvae and other small aquatic insects.

behavior This species perches on rocks or riverbank vegetation. It tends to fly low. The male aggressively defends his territory; he chases intruders by flying in a widening circle. He guards the female while she submerges to deposit eggs on vegetation.

An American rubyspot on forget-me-nots

range Prefers running water. Found mainly on banks of large streams and rivers throughout the U.S. and eastern Canada. Often appears in swarms to feed on emerging mayflies.

How to Tell the Damsels From the Dragons

Though they are quite similar at first glance, damselflies and dragonflies show many differences. Damsels have longer and thinner abdomens, and their eyes bulge out at the sides of the head instead of spreading out in front. Damsels can fold up their narrower wings above their bodies, while dragons hold theirs flat out to the side. Dragons are faster and more direct fliers, whereas damsels tend to flutter.

A damsel has narrower wings than a dragon.

Butterflies

Butterflies add a bucolic presence to watery environments. Several species find the choice of aquatic vegetation and other conditions just right to carry out their four-part life cycle. Butterflies rely on plants for egg laying, for food during the caterpillar stage, for morphing during the chrysalis stage, and for supplying nectar to the winged adults, such that they become important pollinators. Strong, wing-tattering wind can be a butterfly's nemesis, though, as can birds, dragonflies, and damselflies.

Viceroy

Limenitis archippus
Wingspan 2.5–3" **F**

appearance The wings are orange both above and below, with white-marked black edging and black veins. The body is black. This species has oval eggs, a humped and horned olive-brown caterpillar, and a brown chrysalis. Western viceroys are tawny; southern viceroys are browner.

A viceroy butterfly on a red Lucifer plant

food Early-in-year adults eat honeydew, dung, carrion, and fungi. Preferred nectar flowers include aster, goldenrod, and thistle. Caterpillars prefer willow and aspen, and they eat their own eggshells.

behavior Males frequent host plants. Females lay two or three eggs per plant. This species breeds two or three times a year. The last caterpillars of the year make a rolled-leaf shelter for overwintering.

range Found along various wet habitats ranging from ponds and lakes to marshes and rivers.

Atlantis Fritillary

Speyeria atlantis
Wingspan 1.75–2.6" **F**

appearance This varies depending upon region and climate. The species is typically orange-brown above with various black markings and black edges. Below it is orange or reddish brown with markings and silver spots near the wing's edge. The egg is yellow. The small, purplish black caterpillar has light stripes and orange spines. The chrysalis is speckled brown.

food Adults sip nectar of milkweed, mint, burdock, oxeye daisy, and crown vetch, among others. Caterpillars feed on host plants and violets.

Atlantis fritillary

Acadian Hairstreak

Satyrium acadica
Wingspan 1.1–1.25" 🄵

appearance This species is gray-brown above with an orange spot and a small tail on each hind wing. It is silver-gray below with a blue marginal spot, an orange border, and bands of black spots. The caterpillar is green with yellow dashes enclosed by two white lines.

behavior Males seek females along streams, bogs, and wet meadows. Females lay eggs individually on leaf litter near violets. There is one brood per year.

range Found mainly from the Maritime Provinces in Canada south through the Appalachians across to the Great Lakes, south through the Rockies; also eastern Alaska.

food Adults sip nectar from flowers, including milkweed, white clover, and thistles. Caterpillars feed on willows.

An Acadian hairstreak on a milkweed blossom

behavior The male alights on low vegetation near host plants as it looks for females. The female lays eggs on twigs. One brood is produced per year. Eggs overwinter and hatch in spring, and the caterpillars begin to feed on willows.

range Found along willow-lined streams, marshes, meadows, moist woodlands, and seeps, from British Columbia across southern Canada to Nova Scotia and northern U.S.

Viceroy vs. Monarch

These two butterfly species mimic each other in their appearance, a phenomenon called Müllerian mimicry. Both species are toxic and distasteful, and their similarity reminds potential predators of this fact. Some differences exist. Monarchs are bigger, and they tend to flap and glide, while viceroys flutter. Viceroys have a horizontal black band across their hind wings.

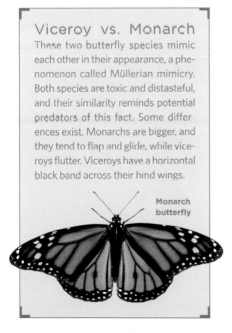

Monarch butterfly

ON LOCATION

Big Bend National Park

- Ecological crossroads
- Distinctive habitats
- Abundant wildlife
- Remote getaway

If you flew over the southern tip of Texas, where the Rio Grande turns to the southeast and then back to the northwest, you'd see a baked and apparently barren landscape cut by canyons, a few thin lines of green along the Rio Grande and smaller streams, and additional green areas on parts of the mountains. The 1,250 square miles of Big Bend National Park, below you, is the nation's largest protected area of the Chihuahuan Desert, North America's largest desert, which sprawls across parts of northern Mexico, Texas, New Mexico, and Arizona.

Desert doesn't mean "dead." Big Bend Park has river's edge environments along the Rio Grande and smaller streams, as well as desert habitats and pinion-juniper-oak woodlands on the mountains. Biologists describe Big Bend as an ecological crossroads where the ranges of eastern and western species overlap.

Climates, Habitats, & Variety
With elevations as high as 7,831 feet at the top of Emory Peak in the Chisos Mountains, canyons as deep as

The Rio Grande winds through San Vicente Canyon in Big Bend National Park, Texas.

1,500 feet, and a hot, dry desert floor, Big Bend's wide variations in temperature and available water account for the park's unusual biodiversity. Life in the park includes more than 1,200 species of plants—60 are cactus species—roughly 3,600 insect species, and more than 600 species of vertebrates, including 75 species of mammals and 450 species of birds.

Butterflies are some of the most abundant insects in the park. They include seven species found only in Big Bend. From late summer through the fall, and again in the spring, migrating monarch butterflies pass through the park. You might not see many of the 75 species of mammals in Big Bend because most move around only at night and in the early morning when it is cool. They include white-tailed and mule deer, gray fox, and the piglike javelina, a South American animal that has moved into the Southwest. Because Big Bend is close to the middle of North America, it is on a major bird migration route. You can see northern birds migrating to and from the south, as well as tropical birds that move into the park to breed in the spring. Some birds regularly seen in the park include Mexican ducks, Lucifer hummingbirds, Mexican jays, Colima warblers, gray vireos, and varied buntings, which are otherwise seen only in Mexico.

Big Bend National Park is also a treasury of fossils from the last 35 million years of the period when dinosaurs wandered the earth. This fossil record continues into the Cenozoic era, or the age of mammals. Paleontologists have found fossils of 90 dinosaur species and nearly 100 plant species in the park, and they continue to find more. The National Park Service warns that it is illegal to take

Rafters float through a Rio Grande canyon.

fossils, as well as living plants and animals, from parks.

A Remote Wilderness

The National Park Service also reminds visitors that Big Bend National Park is in one of the most remote and least populated areas of the United States. One advantage of this remoteness is a night sky with no light pollution; the number and brightness of the stars will dazzle you on the area's many cloud-free nights. It also means, however, that you can't expect to call for help on your cell phone. Summer heat has killed unprepared hikers, and bitter cold can arrive quickly in winter. You can explore much of the park in your car, and you can cover even more on hikes and river rafting trips. Before visiting, you should read the advice and warnings on the National Park Service's Big Bend website. Then, before setting out to explore, talk with the park rangers, who will help make your journey enjoyable and safe.

Bugs & Beetles

All bugs are insects, but not all insects are bugs. True bugs belong to the order Hemiptera, which translates to "half wings." Half-leathery, half-membranous forewings fold flat to the back. Beetles come with a hard-shelled case: tough forewings cover membranous hind wings and usually meet neatly in the middle. Many bug and beetle species are aquatic.

A giant water bug, one of the largest U.S. insects

Giant Water Bug

Lethocerus americanus
L 1.5–4" **F**

appearance The body is flat and brown, and the beak is brown with piercing and sucking mouthparts. The fat front thighs have a groove for folded lower legs. Forewings are leathery.

food Adults eat a wide range of aquatic insects but also catch and consume fish, tadpoles, frogs, and salamanders. An injected anesthetic saliva subdues the prey and liquefies its innards.

behavior This bug breathes through tubes in the rear abdomen. Females lay eggs in neat, attached rows on plants or other above-water support.

range Usually found in vegetation at the bottom of shallow standing water of ponds and pools throughout North America. They also fly to porch lights, hence the nickname electric light bug.

Common Backswimmer

Notonecta undulata
L 0.25–0.38" **F**

appearance The body is black on the underside and white to dark green on the back. Forewings are whitish with red markings. The eyes are large and black. The hairy hindmost legs, used for rowing, are much longer than the grasping forelegs.

food A very active predator, the species eats insects, larvae, snails, small crustaceans, small fish, tadpoles, and frogs.

Back legs propel a common backswimmer.

Whirligig beetles travel on the water's surface.

behavior The female attaches long white eggs to underwater plant stems. The bug flips itself onto its back and uses its hind legs as oars for locomotion just under the water's surface. Backswimmers deliver a painful bite.

range Occurs in ponds and streams and will tolerate shallow, stagnant pools—even swimming pools. Found throughout North America.

Large Whirligig Beetles

Dineutus species
L 0.4–0.6" **F**

appearance The body is wide and oval, dull or shiny black. Each forewing has nine impressed lines. It has slender, long forelegs; short, paddle-like hind legs; and short, clubbed antennae. Its divided compound eyes can see above and below water.

food Adults eat a variety of aquatic insects. Larvae eat mites, snails, and aquatic insects.

behavior Whirligigs walk or glide across water. The antennae help detect location of obstacles and prey. The beetles often congregate in groups. The name comes from a habit of swimming madly in circles when threatened. Females lay eggs in rows or clumps on aquatic plants.

range Found in ponds, lakes, and slow-moving waters throughout North America. Overwinters in mud, debris.

Common Water Strider

Gerris remigis
L 0.5–0.6" **F**

appearance The body is slender, wingless, and brown or black. The legs are very thin, and the front legs are shorter than the middle or hind pair. The water strider has sucking mouthparts.

food It eats aquatic insects such as larvae that rise to surface. It also takes advantage of land insects that fall into water. It grasps prey with its front legs.

A pair of common water striders mate.

behavior The water strider walks on water by not breaking the surface tension, which accounts for the secondary name of pond skater. It communicates at breeding time by making ripples on the water's surface. The female lays neat rows of cylindrical eggs at the water's edge. Usually seen in groups, it can overwinter under leaf litter.

range Found on the surface of quiet waters such as ponds and slow streams throughout North America.

An adult mayfly lives for only a day or less.

Mayflies & Relatives

By the time they become adults, mayflies, caddis flies, and stone flies live on borrowed time. For mayflies, that window may last only from dusk to dawn. Caddis flies and stone flies survive for a couple of months, tops. For these three groups, longevity lies in the aquatic young. The naiad stage may span years, and it displays fascinating adaptations to life in the water.

Mayflies

Various species
L 0.2–1.2" **F**

appearance The adult body ranges in color from white to yellow to reddish to brown. It has large, veined, triangular forewings; small or no hind wings; and long front legs. Two or three long tail filaments emerge from the abdomen. The naiad varies. It has three tail filaments, an often flattened body, and gills on the abdomen.

food Adults have no biting mouthparts and do not eat. Naiads eat algae, microscopic plants, organic debris, and small insects.

behavior Adults live for a day or less; naiads can live for up to four years. Emerging adults swarm, mate, and lay eggs in the water; adults then die.

range Found near or in a variety of watery habitats with muddy or sandy bottoms, including ponds, lakes, streams, and quiet or rushing rivers throughout North America.

A caddis fly—a small, mothlike insect

Fly-fishing lure

Flies for Fishing The larvae
and adults of mayflies, caddis flies,
and stone flies rank high among fish
food favorites. Knowing this, fishers
tie on to their hooks flies that resemble
these insects in order to attract fish.
Dry flies that float on the surface and
drift with the currents mimic adults,
while wet flies that sink stand in for
underwater larvae.

Caddis Flies

Various species
L 0.5-1" **F**

appearance The adult has hairy,
mothlike wings held above the body,
long antennae, and thin legs with
spurs. The color is usually pale, brownish, or grayish. Aquatic larvae, called
caddisworms, often build cases of
sand, twigs, debris, and their own silk.

food The short-lived adult rarely eats.
The larva eats small aquatic insects,
algae, and organic debris. Some larvae
spin silk nets to trap food.

behavior Adults are poor fliers.
After mating, the female may attach
eggs to vegetation or drop strings of
eggs into water.

range Found on foliage or bark near
ponds, marshes, lakes, and quiet rivers
throughout North America.

Stone Flies

Various species
L 0.25-2.5" **F**

appearance An adult's color usually
ranges from yellow to green to gray to
brown. The hind wings fold flat under
forewings and over the back, covering
a long abdomen. Naiads have two tail
filaments and long antennae.

food Adults live for one to four weeks
and rarely eat. Those that do feed eat
pollen and algae. Naiads eat aquatic
insects, algae, and organic debris.

behavior Adults crawl among stones
in or near water—hence their common
name. Stone flies are very poor fliers
and mostly stay close to home. They
often flutter weakly around lights.
Females lay eggs over water.

range Found near clear, running
streams and rivers throughout North
America. Stone fly presence is considered a sign of high-quality oxygen-
ated water.

Stone flies are intolerant of water pollution.

Flies

Flies often dampen our enjoyment of the water's edge. Females of biting species create the misery, as they harvest blood to fuel egg laying. Because flies feed on plant nectar, they are first-class pollinators. Flies have only two wings. The second set, common to other insects, evolved into halteres—small knobs that aid balance and flight maneuvers.

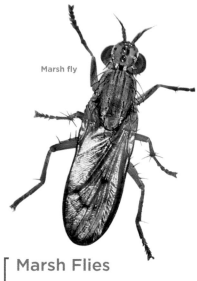

Marsh fly

Marsh Flies

Tetanocera species
L 0.2–0.25" **F**

appearance These small fly species have brownish yellow bodies with amber or yellowish marked wings. The head has prominent red eyes and longish antennae that angle forward noticeably. The upper hind legs display bristles.

food Adults feed on nectar, dew, and other plant liquids. Larvae seem to prey exclusively on (or parasitize) a wide variety of slugs and snails.

behavior Marsh flies often swarm over plants at the water's edge. Adults do not bite animals. These flies are nicknamed slug killers or snail killers due to their predaceous larvae.

range Found on plants along edges of marshes, ponds, lakes, wet meadows, and woodland streams. Common throughout North America, but somewhat more abundant in the north.

Black Flies

Simulium species
L 0.06–0.25" **F**

appearance The body is shiny black to yellow, and the head points downward. A hump behind the head gives the fly a bison-like appearance, which is the source of the common name buffalo gnat.

food Both males and females feed on nectar. Females suck the blood of birds and mammals, especially in forested areas, in daytime. Larvae are filter feeders that extract microorganisms from water.

behavior Black flies often swarm.

Plant stems infested with black flies

Deerfly, also called greenhead fly

They may travel far from their breeding sites in search of food hosts. Larvae use a suction disk on the tip of the abdomen to attach to rocks.

range Near slow-moving water throughout North America but generally rarely seen except when swarming during spring or summer in Canada and northern parts of the U.S.

Deerflies

Chrysops species
L 0.4–0.6" **F** **S**

appearance The body is yellow or black, with yellow-green markings. The body and head are smaller than those of horseflies. The eyes are green or golden with zigzag patterns. The clear wings have dark markings.

food The adult eats nectar, sap, and other plant juices. The female also requires blood for egg development. The larva feeds on organic matter and aquatic insects.

behavior Deerflies, like black flies, have silent flight and a powerful bite. They swarm, mainly around the head and neck. Females lay shiny black eggs on plants or rocks over water.

range Found throughout North America in a wide variety of habitats near water.

safety tip To avoid painful bites from deerflies and horseflies, cover up with light-colored clothing and consider a mesh head net and mesh jacket.

Horseflies

Tabanus species
L 0.5–1.2" **F** **S**

appearance The body is blackish or brownish, and the large head has large, dark eyes, some metallic. Some species have markings on the body. Antennae have three parts.

food Horseflies sip nectar, honeydew, and other plant liquids. The female inflicts a painful bite before a blood meal and injects an anticoagulant to keep the blood flowing. Larvae feed on aquatic insects.

behavior Females swarm around potential victims. Repeated biting can debilitate an animal. Females lay eggs on waterside plants, and the larvae drop into the water. Horseflies transmit some diseases to humans.

range *Tabanus* species are found all over North America in moist settings.

Biting horsefly

Mosquitoes & Midges

As waterside pests, mosquitoes and midges live up to their fly credentials. These small flies swarm and annoy, although in their favor, midges do not bite. Female mosquitoes do—with a vengeance—to obtain blood protein for egg production. Both groups feed on a wide range of animals, including birds, bats, amphibians, and other insects. Immature stages filter water, eat algae, and fall victim to fish and other aquatic creatures.

A dead mosquito lies on an insectivorous round-leaved sundew plant.

Mosquitoes

Various species
L 0.2–0.25"   F S

appearance Mosquitoes are small, slender flies with long, thin legs. Females are usually larger than males. Both sexes have a long mouthpart called a proboscis. The male has feathery antennae, while the female's are sparsely haired. The active aquatic larvae are known as wrigglers.

food Both sexes sip nectar and other plant liquids. Females bite warm-blooded animals. Larvae eat algae and bacteria.

behavior Females lay black eggs singly or in "rafts" on the water. The eggs hatch into larvae, which feed head down in water. Adults emerge from comma-shaped pupae at the water's surface.

range Found near natural and artificial water sources throughout North America. Various species responsible for disease transmission.

safety tip

No water source is too shallow or temporary to thwart an egg-laying female mosquito. Be sure to empty plant saucers and similar vessels.

did you know?

the presence of some midge species as fossils helps scientists learn about environmental and climate change? Midges indicate water quality, giving clues to past conditions.

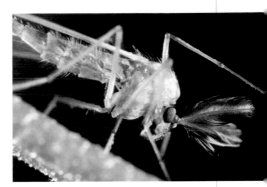

A close-up of a midge

Midges

Various species
L 0.1–0.25" F S

appearance Smaller and more delicate than mosquitoes, midges lack a long proboscis and scales on their wings. Their bodies are brown to green. Midges hold their wings to the side when at rest and show a flattened profile. The male's antennae are very feathery. Midges swarm but do not bite, although some other insects commonly called midges (or punkies) do bite.

food Adult midges typically live for a day or so and do not feed. Larvae feed in the thick layer of detritus, algae, and bacteria on silty bottoms.

behavior Adults emerge from post-larval pupae and burst through the water's surface. They swarm to mingle and then drop below the swarm to mate. Females immediately lay eggs in water or on aquatic vegetation. Eggs hatch into larvae, which settle to the bottom to feed.

range Found near natural and artificial bodies of water in both coastal and inland areas throughout North America. Different species swarm at different times of day.

This illustration depicts the life cycle of salt marsh mosquitoes, from egg to larva to pupa to adult.

Spiders & Horseshoe Crabs

Seldom associated with aquatic environments, spiders do have a solid presence there. Some species utilize the same watch-and-wait strategy as their terrestrial kin, except they use vibrations on the water's surface instead of a rustling web to detect prey. The horseshoe crab (which is not a crab but a relative of spiders) resembles its forebears of 200 million years ago, living in the sea and crawling onto beaches to mate in spring. A female may drag along a courting male while she excavates a hole and lays eggs, which he then covers with sperm.

A fishing spider with prey

Fishing Spiders

Dolomedes species
L up to 2.5" **F**

appearance The fishing spider is large and hairy. Its body is mottled with brown, gray, white, or black, and it closely resembles a wolf spider in size and pattern. It has eight eyes in two rows. The female is bigger than the male—in some species almost twice the size.

food An opportunistic feeder, the fishing spider will eat anything it can capture. It takes water striders and other aquatic insects, small fish, and frogs.

behavior It waits on rocks or vegetation in water or submerges itself in a bubble of air. When prey ripples the surface, the spider attacks and grabs it with powerful front legs. After mating, the female carries the egg sac until the hatchlings hatch.

range Found near ponds, streams, and bogs and in wooded areas throughout much of North America. Largely absent on the West Coast and in western Canada. Some species found on tree limbs.

Horseshoe
crab

Horseshoe Crab

Limulus polyphemus
L 16–20" **S**

appearance This animal has a hard, horseshoe-shaped head-and-thorax section hinged to its abdomen. There are two pairs of eyes, one simple and one compound. Its paired limbs are for feeding, locomotion, and reproduction. A spiky telson, or tail, helps to right a flipped-over crab.

food The horseshoe crab feeds on clams, other mollusks, and worms that it excavates from sand. Its legs move food to a crushing mechanism called gnathobases before moving it on to the mouth. Larvae do not feed.

behavior The animal reaches sexual maturity in about 10 years and can live for up to 20 years. In spring and summer males patrol bay beaches, awaiting the larger females. Females lay 2,000 to 20,000 eggs, which multiple males can fertilize. The eggs hatch into larvae that move into the water.

range Inhabits tidal flats and deeper waters of Atlantic coast from Maine south and along Gulf Coast. Concentration is in Delaware Bay.

did you know?

migrating red knots feed on horseshoe crab eggs as these shorebirds make their 9,000-mile trip from Chile to their Canadian Arctic breeding grounds? Knots are one of several species that rely on the spring bounty.

A group of horseshoe crabs spawn at low tide on an East Coast beach.

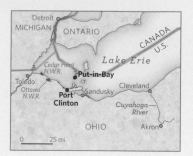

ON LOCATION

Port Clinton, Put-in-Bay, Ohio

- Lakeside relaxation
- Bird-watching
- Freshwater fishing
- Remnants of huge swamp

From time to time each summer, batches of mayflies emerge from the year or two they spent as naiads/larvae on the bottom of Lake Erie for brief flings as flying, mating insects. They all die in a day or so, after some females lay eggs that sink to the bottom of the water and continue the cycle. Mayflies do not bite, sting, or eat, but they splatter on windshields and into our faces and clothing. Dead mayflies make roads slippery and smell like rotten fish. Nevertheless, the mayflies are a sign that, unlike in the 1960s and 1970s, we can enjoy exploring nature in the Lake Erie recreational region around Port Clinton and Put-in-Bay. In the 1960s, Lake Erie, which often had rotting dead fish and algae covering its shores, was a poster child for water pollution.

The Lake's Decline and Recovery

Lake Erie's reputation for being "dead" spread around the world in the days after June 23, 1969, when oil and other waste in the Cuyahoga River caught fire in Cleveland and sent up flames as high as a five-story building. Headlines and television commentators described "the river that caught fire." The 85-mile-long Cuyahoga, which flows into Lake Erie in Cleveland, could not be blamed for Erie's woes, however; 80 percent of Erie's water comes from the three larger upper Great Lakes.

For years, water from cities, towns, industries, and farms around Lake Erie had been feeding pollutants into the lake, and this had led to huge algae blooms. Bacteria eating the dead algae took oxygen from the water, thus suffocating other aquatic life, some of which washed up on the shores dead. Beginning in 1972, the United States and Canada agreed on a series of measures to reduce lake pollution, as both nations were also adopting measures for general cleanups of their rivers and lakes. Mayfly swarms had begun decreasing in the late 1950s, but this early sign of trouble did not have the same impact as rotting fish or a burning river. Mayflies began returning in the 1990s, which is one sign that the lake's aquatic life

Boating at Put-in-Bay on Lake Erie, Ohio

A kayaker sets off from South Bass Island. Each year millions of people visit Lake Erie.

is generally doing well—at least so far. Unfortunately, levels of phosphorus in Lake Erie are increasing, and algae blooms are occurring. Fish die-offs are occurring again in parts of the lake, and authorities have issued advisories against swimming from time to time.

Enjoying Lake Erie Today

Port Clinton and the Lake Erie Islands are known as places to enjoy boating and fishing. There are also opportunities for bird-watching and otherwise enjoying the natural world in a relaxed way, rather than going on a wilderness trek. Lake Erie has more fish and a greater variety of species than the other Great Lakes. These include walleye, yellow perch, smallmouth bass, white bass, and channel catfish.

South Bass Island, where Put-in-Bay is located, has the National Park Service's Perry's Victory and International Peace Memorial, which celebrates Commodore Oliver Hazard Perry's defeat of British ships on September 10, 1813, during the War of 1812, as well as long-lasting peace among the United States, Canada, and the British. In June 2011 the National Park Service held its third annual Return of the Mayfly festival there to educate visitors on "those insects we all love to hate."

Several state parks on the islands and mainland have camping areas and nature walks. On the mainland, the Ottawa, Cedar Point, and West Sister Island National Wildlife Refuges protect roughly 9,000 acres of habitat, including the last remnants of the Great Black Swamp. The region is important to migratory waterfowl, songbirds, and shorebirds. Until the 1830s, the Great Black Swamp covered all or part of 12 counties in northwest Ohio as well as part of Indiana. It was mostly drained and settled between the 1820s and 1900.

As the sea travels over the ribbed bottom of the flats, fish move in from the deep, dissolving in and out of sight around my feet. Though they seem mute, they are not. They fill the waters with belches and cries, calls of courtship, alarm, aggression, and fright.

—JENNIFER ACKERMAN, *NOTES FROM THE SHORE*

[4]

FISH

Many stingray species are found in shallow waters with sandy bottoms, where they are often easy to see when disturbed by a boat.

Sharks & Rays

Sharks, rays, and skates belong to a class of fish with skeletons made of cartilage, not bones. Only their teeth, and sometimes their vertebrae, consist of bonelike material. Sharks, of course, are famous predators. Stingrays are known for their dangerous stings, but a group of rays called eagle rays do not sting. Skates look like short-tailed rays. Most are harmless, but some species produce powerful electric currents around their bodies.

Stingrays like this one whip their long tails around to embed stingers in enemies.

Stingray

suborder Myliobatoidei
L up to 6.5' **F** **S**

appearance Most stingrays are tan or grayish above and white below, with flat bodies, winglike pectoral fins, and a long, whiplike tail. The eyes are on top of the body, and the mouth is underneath. Stingrays hide just under the surface of a sandy sea bottom and may be invisible except for their eyes.

food Stingrays eat mollusks, crustaceans, and worms; they also scavenge on carrion.

behavior Stingrays dig food from the seabed and use strong teeth to crush shells. Little is known about their breeding habits. Eggs hatch inside the female's body and are kept alive by nutrients in yolk sacs. Later they absorb food through the mother's uterine wall until they are born.

range Tropical, subtropical, and some temperate-zone oceans worldwide.

safety tip

If you step on a stingray, its tail will most likely whip up and sting you. Shuffling your feet and splashing as you wade will scare a stingray away.

Spiny dogfish

did you know ? **Atlantic stingrays** found upstream in Florida's St. Johns River are the only known permanent freshwater population of these rays?

Spiny Dogfish

Squalus acanthias
L up to 42" **S**

appearance This small shark species is gray to brown above, with white spots on the sides, and a light gray to pure white belly. On its back, it has two dorsal fins, the first of which is larger than the second. On the front of each fin, a sharp spine extends upward from the base.

food The diet includes various fish, smaller sharks, octopuses, squid, crabs, and shark egg cases.

behavior Schools of hundreds to thousands of spiny dogfish, often of the same size or sex, gather offshore, where mating takes place. Young have a two-year gestation period—perhaps the longest period of any vertebrate. During that period they have cartilage-like sheaths on the spines so that they will not harm the mother's organs.

range Occurs mostly in shallow water, but sometimes farther offshore, on Atlantic coast from Labrador to Cuba; uncommon south of North Carolina.

Skates

family Rajidae
L up to 9' **S**

appearance Skates resemble rays, but they have a diamond shape and a shorter, thicker tail with no spiny stinger. Many of the 200 species have thorny projections on the upperparts. They come in various colors, usually pale or dark brown, and many have dark spots on the upper side.

food Skates eat bottom-dwelling fishes, shrimp, worms, and clams.

behavior Like rays, skates spend time motionless and camouflaged on the sea bottom. Their young hatch from eggs in an oblong case with a thin spine at each corner. Growth and reproductive rates are slow.

range Global oceans from the Arctic to the Antarctic, from shallow coastal shelves to great depths.

Skate

Saltwater Fish

All fish are cold-blooded vertebrates with fins, gills, and a sense organ unique to fish. This organ is a row of tiny holes in the skin from behind the gills to the tail, and it is sensitive to vibrations. Fish use it to swim in schools, to find prey, and to avoid predators. Any creature with these characteristics, including a seahorse, is a fish. Some fish are adapted to live in salt water, while others thrive in fresh water, and some in both.

Seahorse

genus Hippocampus
L up to 12" **S**

appearance This is a true fish with a profile resembling a horse, yet it has no scales. Its colors range from mottled brown to bright orange, and it can change its color to match its surroundings.

food Seahorses feed on small shrimp, small fish, crustaceans, and plankton.

behavior A seahorse is a member of a large family called pipefish. It swims upright and anchors itself by wrapping its tail around objects. Its mouth is a long tube that sucks up tiny organisms. The male holds fertilized eggs in a pouch on the front of his body for 9 to 45 days, until fully developed miniature sea horses emerge. Adults do not nurture their young.

range Sea grass beds, coral reefs, mangroves worldwide in tropical and temperate waters.

Sculpin

superfamily Cottidae
L up to 12" **F S**

appearance Sculpins are strange fish, with oversize heads and often with tough spines. Most are mottled in dull brown or red, depending on their surroundings.

Seahorse

safety tip

Spines on the fins of sculpin have venom glands that cause extremely painful wounds, with redness and swelling. You should not pick up and handle a sculpin that you may see in a tide pool or a freshwater stream.

Sculpins never win fish beauty contests.

food Sculpins eat small fish, fish eggs, crab, squid, and octopus.

behavior More than 300 species worldwide are saltwater or freshwater bottom dwellers. Some species in West Coast streams migrate to estuaries to spawn. Oddly for a fish, some species in tide pools can breathe air and survive for a while out of water.

range The diverse habitats include deep oceans, tide pools, clear freshwater streams, and even the bottoms of the Great Lakes. Marine species occur in all Northern Hemisphere oceans from Arctic to tropics.

Pipefish

subfamily Syngnathidae
L up to 16" **F** **S**

appearance Pipefish are odd looking, with a long, thin body and tail. The name refers to the mouth's straight, round, pipelike shape. Most species are dull green or olive, and some change color to blend with their surroundings.

food The diet includes young fish and small crustaceans.

behavior Some pipefish have elaborate courtship rituals. The males of some species brood eggs. Young are born free-swimming and independent of parents, who may view them as food. Many are weak swimmers in open water; they move slowly with rapid flapping of their dorsal fins.

range Globally in tropical, temperate oceans and fresh water; may be abundant in protected areas such as sea grass beds, sandy lagoons, or coral reefs.

Pipefish (at right, above a shrimp) are sometimes found in sea grass beds.

Striped bass

Striped Bass

Morone saxatilis
L up to 6.6' **F** **S**

appearance These bass have streamlined, silvery bodies with dark horizontal stripes on the sides. Fishing enthusiasts often call them stripers.

food Striped bass feed on anchovies, spot, menhaden, herring, shad, white perch, and yellow perch.

behavior In winter and spring, mature Atlantic coast fish move into tidal freshwater to spawn. The young remain in these rivers and estuaries until age four to eight. The adults return to the ocean after spawning and migrate to New England waters. In fall, they migrate southward off North Carolina and Virginia coasts. Many freshwater populations are maintained by stock from hatcheries.

range Native to the North American Atlantic coast from the St. Lawrence River to the Gulf of Mexico; introduced to North America's Pacific coast and to many inland lakes.

did you know ? **fish and many other** aquatic animals reproduce by spawning? They release eggs into the water, where some eggs are fertilized. The eggs have nutrients to sustain the embryos as they develop into young.

Red Snapper

Lutjanus campechanus
L up to 39" **S**

appearance This fish is easy to recognize by its light red color, flattened oval shape, spiny top fin, large head, small red eyes, and somewhat pointed snout.

food Juveniles feed on zooplankton. Adults feed on shrimp, squid, octopus, various small fishes, crustaceans, and mollusks.

A school of snappers

behavior Adults inhabit rocky bottoms, ledges, ridges, and artificial reefs such as oil rigs and shipwrecks. They form large schools, usually consisting of fish of similar size. After spawning, eggs float on the ocean surface and hatch 20 to 27 hours after fertilization. Small larvae are briefly planktonic.

range Waters 30 to 200 feet deep in the Gulf of Mexico, southeastern U.S. Atlantic coast; rare north of Carolinas, but sometimes seen as far north as Massachusetts.

Tarpon

Megalops atlanticus
L up to 8' **F** **S**

appearance The sides are bright, shiny silver, and the back is darker greenish or bluish. The tarpon has unusually large scales, and its large mouth is turned upward. A long,

Anglers consider tarpon a great game fish.

thin, pointed portion of the dorsal fin arches backward.

food First-stage larvae absorb nutrients from seawater; later stages and juveniles eat zooplankton, insects, and small fish. Adults hunt at night and prey on fish, shrimp, and crabs.

behavior Tarpons typically spawn in May, June, and July in offshore spawning areas where currents move larvae to inshore nurseries. Larvae have three stages. Adults inhabit coastal waters, bays, estuaries, and mangrove-lined lagoons in the tropics and subtropics. They often travel up rivers into freshwater, and they can tolerate oxygen-poor environments because a modified air bladder allows them to breathe air.

range Occurs in warm waters along the Atlantic coast from Virginia to Brazil, Caribbean coasts, and around the Gulf of Mexico.

Eat the Right Fish

We hear that eating more fish is healthy. We also hear about toxins such as mercury in some fish and read that many species are declining because of overfishing. How do you know which fish are healthy and safe, and how do you avoid hurting a declining species' population by eating the fish? The Monterey Bay Aquarium's "Seafood Watch" can help. Its website, booklets, and applications for smart phones and tablets enable you to check out particular species. The "Fish Watch" website of the federal wildlife and oceanic agencies can also help. Search for "Seafood Watch" and "Fish Watch."

A youngster admires his catch.

Bluefish are a favorite of anglers on and near shore.

Bluefish

Pomatomus saltatrix
L up to 3.5' S

appearance Bluefish are grayish blue-green on top, fading to white on the lower sides and belly; they have broad forked tails.

food These fish eat worms, crustaceans, and smaller fish.

behavior Bluefish migrate south in fall and north in spring in large schools, following small fish often near the shore; they are most abundant off Florida in winter. Juveniles prefer sandy bottoms but also inhabit locations with mud, silt, or clay bottoms near the shore and in estuaries. Bluefish are extremely aggressive; they chase large schools of small fish through the surf zone. Their populations are highly cyclical, with abundance varying widely over spans of ten years or more.

range Nearly worldwide in temperate and tropical oceans and estuaries.

safety tip

Watch for bluefish pursuing small fish into the surf, churning the water, and biting everything they encounter—including people.

Flounder

order Pleuronectiformes
L up to 24" S

appearance A flounder's body is flat, oval, and short tailed. Both eyes are on the same side of the head. The upper side is usually tan or brown, plain on sandy sea floors, or mottled to match seafloor habitats. The underside is white.

food Flounder eat crustaceans, other invertebrates, and fish.

behavior These fish most often lie camouflaged on seafloor sediments and ambush their prey. Some leave the sea bottom to hunt. In an amazing developmental process, the larvae have an eye on each side of their heads; then, during their metamorphosis to

Flounders belong to a group called flatfish.

adulthood, one eye migrates to the other side.

range Worldwide among various species, mostly in shallow waters.

Atlantic Mackerel

Scomber scombrus
L up to 16.5" **S**

appearance This species is iridescent blue-green above, with a silvery white underbelly, 20 to 30 wavy black bars across the top half of its body, and a narrow, dark streak along each side.

food The Atlantic mackerel eats crustaceans such as krill and shrimp.

behavior This fish migrates south to spawn in summer. It forms large schools near the surface, where it is an abundant and important catch for major commercial fisheries and

Atlantic mackerel

a favorite of anglers. It overwinters in deeper waters, then moves closer to shore during its migration north in spring, when water temperatures range between 52°F and 57°F. It spawns offshore between Massachusetts and North Carolina between April and May and in the Gulf of St. Lawrence in June and July.

range Occurs on both sides of the Atlantic Ocean; in North America, from Labrador to North Carolina.

Eels are long-distance migrants.

American Eel

Anguilla rostrata
L up to 4' **F** **S**

appearance This eel has a snakelike body and a small, pointed head; in fresh water, it is brown on top and yellowish tan below.

food In fresh water it eats dead fish, invertebrates, carrion, and insects.

behavior The young hatch in the Sargasso Sea; larvae mature as they migrate to North America. Eels reach sexual maturity in 8 to 24 years and then move back toward the Sargasso Sea, where they metamorphose into a silver eel phase. During migration, their eyes double in size and increase in their sensitivity to blue, which enhances deep-water vision. Adults die after spawning.

range Coasts and streams from western Greenland and eastern Newfoundland south to the Gulf of Mexico, Panama, and the Caribbean.

ON LOCATION

John Pennekamp State Park, Florida

- An undersea state park
- Highly accessible coral reefs
- Protected habitats
- Delicious invasive fish

The Florida Keys allow us to explore a unique water's edge: the bottom of a warm ocean. John Pennekamp State Park on Key Largo would be a good first stop, especially if you are planning to dive or snorkel. You could spend a day or so talking with the park's rangers and volunteers to get advice on the best ways to enjoy the Florida Keys National Marine Sanctuary, which includes the ocean surrounding the islands. You can also experience the keys' underwater, beach, and land environments. The park offers close-up views of coral reefs and their plants, fish, and other creatures in its 92 square miles of ocean for scuba divers, snorkelers, or—for those who want to stay dry—passengers in glass-bottom boats. On land the park has boardwalks along a mangrove-lined estuary and through a dense, dark tropical hammock. You can go swimming at the beaches.

Boats are moored over reefs at the underwater John Pennekamp State Park in Florida.

Protected Water's Edges

The Dagny Johnson Key Largo Hammock Botanical State Park, a mile north of Pennekamp State Park, is also worth visiting. It has six miles of nature trails through the largest remaining subtropical West Indian hardwood hammock forest in the continental United States, and it is the habitat of more than 80 protected species of plants and animals.

The park is named after the environmental activist who led the fight to stop a large condo development on Key Largo. Environmentalists fought the development because they feared that pollution from so many additional residents would kill the eastern end of the coral reef running the length of the keys roughly five miles offshore. It is the only living coral barrier reef in the continental United States.

Many tourists visit the Florida Keys to snorkel the shallower reefs or to explore the deeper reefs if they are experienced scuba divers. Since 1990 the Florida Keys National Marine Sanctuary, which includes 3,900 square miles of water around the keys from Biscayne Bay off Miami to the Dry Tortugas, has restricted potentially harmful activities. Its southern boundary follows the 300-feet deep contour, while the shore boundary is the mean high-water mark. This means that when you go wading in the ocean around the keys, you are in the sanctuary. In addition to encompassing the world's third largest barrier reef, the sanctuary includes extensive sea grass beds, mangrove-fringed islands, more than 6,000 species of marine life, and historical artifacts such as shipwrecks. Protective regulations include reducing the impact of divers and snorkelers on

A scuba diver explores a coral reef.

the reef and other parts of the ecosystem. The National Oceanic and Atmospheric Administration and the State of Florida, which manage the sanctuary, encourage divers and snorkelers to book tours with Blue Star operators who have made a commitment to reduce reef damage.

Eat Invasive Fish, Save a Reef

If you see lionfish on the menu in the Florida Keys, order it. You will help reduce the population of an invasive fish that threatens reefs. Lionfish derbies, in which scuba divers complete for cash and other prizes for catching the most or the largest lionfish, are becoming popular. These fish have already damaged reefs in the Bahamas and elsewhere. They are native to the Indo-Pacific Ocean and the Red Sea and were probably released in the western Atlantic by saltwater aquarium owners. Lionfish are a concern in the keys because they multiply rapidly and eat grouper and snapper, which are important commercial fish, as well as parrotfish, which graze on algae that kill corals. More restaurants are serving lionfish, which many people think are delicious.

No matter what you plan to do in the Florida Keys, the best time to visit is during the (relatively) dry season in December through April when the humidity is comfortable.

Freshwater Fish

Most, but not all, freshwater fish never visit oceans. Anadromous fish, such as salmon, shad, and some trout species, spawn in fresh water but otherwise live in oceans. Catadromous fish are the opposite: They live in freshwater and spawn in the ocean. Eels are the only catadromous North American fish.

Minnows are a key of freshwater food webs.

Minnows

family Cyprinidae
L up to 14" **F**

appearance Minnows are generally, but not always, very small and silvery. Some have accents of yellow, orange, or blue.

food Most minnows eat algae, insects, and microorganisms.

behavior Minnows belong to the largest North American fish family, with 2,400 species, including carp, shiners, dace, and others. A few minnow species build nests by piling up gravel from the streambed or simply excavating it. Spawning occurs in the nest. The male covers the eggs and may stay to guard them. In food webs, minnows are important links between small organisms and larger predators such as fish, birds, and mammals.

range Almost all freshwater habitats in North America.

Common Carp

Cyprinus carpio
L up to 47" **F**

appearance Carp are silvery, with large, meshlike, thick scales. Barbels (whiskerlike projections) around the mouth aid carp in hunting food in murky water.

food Carp eat stream- and lake-bottom midges, other larval insects, snails, small crustaceans, slugs, vegetation, detritus, and plankton.

behavior Habitats include lakes, ponds, and rivers. Carp tolerate a range of conditions, from pristine to polluted waters.

Carp live on bottoms of rivers and lakes.

Spawning occurs along shores and in backwaters. Carp may burrow into embankments in search of food.

range Native to Europe and Asia; now in most of North America, where it was widely introduced from 1880s to 1930s by U.S. and state wildlife agencies to replace declining native fish.

American shad

did you **know** ?

shad were introduced into San Francisco Bay and the Sacramento River in the 1880s? They are established in several Pacific coast rivers.

American Shad

Alosa sapidissima
L up to 30" **F S**

appearance Its thin, silvery body varies from greenish to dark blue on its back. The tail is forked. Adults have a large, dark shoulder spot followed by several smaller, paler spots.

food The shad eats mainly plankton but sometimes feeds on small shrimp, fish eggs, and small fish.

behavior A member of the herring family, the American shad spends most of its life at sea. It swims up freshwater rivers and into bays to spawn in spring and early summer, when the water temperature is between 50° and 55°F. After spawning, shad north of Cape Hatteras move offshore to feed. They may return to where they hatched to spawn in future years. Most shad spend four years at sea before returning to spawn for the first time. Shad further south usually die after spawning.

range Abundant along the Atlantic coast from southern Labrador to northern Florida.

Make a Minnow Trap You can make a trap with two 2-liter soda bottles to capture minnows for bait or for a freshwater aquarium. Cut off the bottom few inches of one bottle. Cut more off the bottom of the other (as shown at right). Keep the cap on the larger bottle; recycle the other cap. Push the smaller bottle into the other one until the bottom edges line up. With the bottles together, carefully punch holes around the bottom edges and run string in and out of them to hold the trap together. Tie a long cord around the top and the other end to something on the bank. Then put your trap into a stream with the open end facing upstream. Most minnows that swim into the open end will not find their way back out.

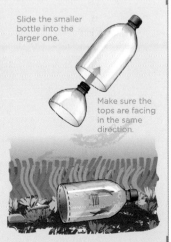

Slide the smaller bottle into the larger one.

Make sure the tops are facing in the same direction.

Bluegill

Lepomis macrochirus
L up to 16" **F**

appearance The bluegill's body is saucer shaped and olive green over-all. The sides of the head and chin are dark blue, and it has a yellowish breast. The fish usually has vertical bars on its sides, but these are not always distinct. The breast of a breeding male is bright orange.

food This fish eats young microscopic and near-microscopic animals, insect larvae, worms, crayfish, leeches, snails, and small fish; it will eat aquatic vegetation if food is scarce.

behavior The bluegill favors shallow water with weed beds, submerged logs, and drop-offs. It moves to deeper water in the summer. It is usually found in schools of 10 to 20. The male scrapes out a 6-to-12-inch spawning nest from sand or gravel, and this is where the eggs are laid. After spawning, the male guards the nest by chasing away larger fish.

range Occurs naturally east of the Rocky Mountains; introduced into most other regions of North America.

look for bluegills or largemouth bass males guarding or defending their nests on the bottom when you snorkel or dive in freshwater lakes.

Largemouth Bass

Micropterus salmoides
L up to 29" **F**

appearance The body looks heavy; it is olive green with dark blotches in lines along the flanks. The upper jaw extends beyond the rear edge of the eye, thus distinguishing it from small-mouth bass.

food Largemouth bass eat small fish, shrimp, insects, black fly larvae, worms, mussels, and frogs.

Bluegills make nests and defend their eggs.

Anglers consider largemouth bass one of the best game fish.

behavior This fish prefers clear water with abundant aquatic weeds. It spawns in water two to six feet deep with a sand or gravel bottom and uses its tail to fan out a nest for eggs. The male fertilizes eggs and then guards the nest, eggs, and young by attacking anything that approaches. After hatching, the fry swim in tight schools until they reach one inch in length.

range Lakes, ponds, rivers, and large creeks in most of the U.S.

Channel Catfish

Ictalurus punctatus
L up to 4.5' **F**

appearance This fish has an elongated body with no scales; it is olive brown to slate blue on the back and sides. A deeply forked tail distinguishes this species from most other catfish.

food Young eat mostly insects, crayfish, and other fish; adults eat crayfish, mussels, clams, small fish, insects, wild vegetation, and carrion.

behavior Channel catfish live in a variety of freshwater habitats. They migrate to spawning and feeding areas in shallow waters in late spring and early summer. The male builds a nest in an underwater hole, in a hollow log, or among submerged rocks. Eggs hatch five to ten days after spawning. The fish migrates to deepwater habitats in winter.

range From Hudson Bay region south to Florida and northern Mexico, and from East Coast to the Rockies.

Barbels (whiskers) give catfish their name.

did you know ?

the barbels around a catfish's mouth are taste buds? Along with thousands of taste buds on the skin and a keen sense of smell, catfish can locate food far away or in murky water.

Freshwater trout spawn in fast-moving streams.

Rainbow/Steelhead Trout

Oncorhynchus mykiss
L 45" **F** **S**

appearance The torpedo-shaped body is generally blue-green or yellow-green with a pink streak along the side and small black spots on the body and fins. The underbelly is white. The male steelhead (named for its steel-gray head color) varies from rich reddish on its sides during spawning season to silvery in its ocean phase.

food The young eat aquatic insects, including caddis flies, mayflies, midges, and water fleas. Adults add small fish, snails, leeches, fish eggs, and algae to their diet.

behavior Rainbow and steelhead trout are the same species. Individuals develop differently depending on their environment. All hatch in gravel-bottomed, fast-flowing rivers and streams. Rainbow trout live entirely in fresh water. Steelhead trout migrate to the ocean, usually grow larger than rainbow trout, and return to the river where they hatched to mate.

range This species is native to North American rivers flowing into the Pacific Ocean, but it has been introduced widely across North America and around the world as a tasty catch for fly-fishers.

did you know? **all North American** freshwater fish that are called trout are members of the salmon family, Salmonidae? Genetic tests have shown that some trout with different appearances and names are really the same species.

Rainbow trout

Chinook Salmon

Oncorhynchus tshawytscha
L up to 58" **F** **S**

appearance The chinook has two forms. When at sea, it has silver flanks, a blue-green back and top of the head, and black spots on the upper half of the body. During its spawning run, it has a red or partly red body. This is the largest salmon species.

Chinook salmon in its spawning red color

food The young eat insects and various crustaceans; adults eat mostly other fish.

behavior The female digs out a shallow nest in an inland stream and lays pea-size eggs that males fertilize. Then she covers eggs with gravel. Adults guard the eggs for a few days to a month and then die. The young spend six months to three years in streams and brackish estuaries, where their body chemistry changes to tolerate life in salt water. They spend one to eight years in the ocean before returning to their home stream to spawn.

range natural distribution from Alaska to southern California. They were first introduced into the Great Lakes in the 1870s, failed to maintain a population, and were reintroduced in the 1960s; now maintained by stocking young fish.

look for salmon battling white water and leaping over waterfalls on their way upstream in the Pacific Northwest to spawn where they were born.

Snakehead Invasion With a somewhat snakelike head, an ability to travel overland, rapid population growth, no natural enemies, and a taste for aquatic creatures including fish, invasive snakeheads are an environmental horror story. These freshwater fish, native to Asia and Africa, established a breeding population in the Potomac River by 2004, and they might be breeding elsewhere as well. Poisons that kill snakehead also kill other fish. People who dumped pet snakeheads or wanted to start a population of the tasty fish may be responsible for introducing them. Federal law prohibits importing snakeheads or taking them across state lines.

With their large mouths, snakeheads are able to eat many native fish.

RECOMMENDED
DR. BEACH
RECOMMENDED

ON LOCATION

Firehole River, Yellowstone National Park

- World-class trout fishing
- Pristine wilderness
- Geysers and hot springs
- Stunning waterfalls

Cliff Geyser erupts near the Firehole River.

If you are an angler of the fly-fishing persuasion, Yellowstone National Park's Firehole River should be one of your do-not-miss destinations. You would not have to worry about family or traveling companions who would rather meet fish on a plate in a fine restaurant than catch one while wading in cold water. While you try to convince a trout that your carefully tied fly is one of the river's tasty mayflies, your companions can enjoy breakfast or a fish lunch at the Old Faithful Inn—the Firehole flows past the lodge, as well as Old Faithful itself. They can also explore the geysers while you fish. If you make reservations well ahead of time, you can stay at the lodge or in one of the nearby cabins.

Glorious Water's Edges

With the Firehole as well as other rivers, creeks, mountain streams, lakes, and ponds, Yellowstone offers endless water's-edge explorations. The park's must-see water's-edge sites include the Yellowstone River, where it plunges over the 109-foot Upper Yellowstone Falls and then a quarter mile downstream over the 94-foot Lower Yellowstone Falls into the 24-mile-long Grand Canyon of the Yellowstone, which is between 800 and 1,200 feet deep.

A fly-fisherman casts his line in a Yellowstone river while bison bathe and graze in the background.

The Firehole contains three major waterfalls. You can view the 150-foot Kepler Cascades from a turnout off the Grand Loop Road two miles south of Old Faithful Village. The 40-foot Firehole Falls are a half mile from where the Firehole and Gibbon Rivers meet at Madison Junction. One of the best views is the Cascades of the Firehole in Firehole Canyon. Take the narrow, paved two-mile Firehole Canyon Drive through the canyon after turning off the Grand Loop Road about a mile south of where it meets West Entrance Road at Madison Junction. On the canyon road you will pass a swimming hole, one of the two places where it is legal and safe to swim in Yellowstone's thermally heated water. In the canyon you will probably see some of the park's abundant aquatic insects, ospreys and other raptors, smaller birds, bison, elk, and possibly river otters, which are the top predators along the park's waters.

Life in Hot Water

Old Faithful and the park's other thermal features, where heat from deep in the earth forces boiling water and steam to the surface, are also must-see features. They include the Grand Prismatic Spring, which has vivid colors of microbial mats around the edges. In 1966 scientists discovered that the mats are composed of living creatures thriving in temperatures higher than 130°F. The intense blue in the spring's center is the color of deep, pure water. Like the Grand Prismatic Spring's water—but unlike most other waterways in North America—Yellowstone's waterways have remained mostly untouched by water pollution, diversions, or dams. A few days spent trout fishing or enjoying the varied wildlife along the edges of Yellowstone's waters will give you a taste of an almost vanished world. Yellowstone offers you a nice sample of Rocky Mountains waters' edges.

Reptiles are abhorrent because of their cold body, pale color, cartilaginous skeleton, filthy skin, fierce aspect, calculating eye, offensive smell, harsh voice, squalid habitation, and terrible venom.

—CAROLUS VON LINNAEUS (1707–78), FATHER OF BIOLOGICAL CLASSIFICATION

[5]

REPTILES & AMPHIBIANS

An American alligator in algae-covered water

Alligators & Crocodiles

By the 1950s, habitat destruction and hunting had almost driven American alligators and crocodiles to extinction. Federal laws adopted in the 1960s and 1970s prohibited hunting either species. The U.S. Fish and Wildlife Service removed alligators from the endangered species list in 1987. They are now numerous across their range, and federal law allows regulated hunting of alligators. The Fish and Wildlife Service removed crocodiles from the endangered list in 2007 and now lists them as threatened in southernmost Florida (their only U.S. home) and endangered in the rest of their range in Mexico and Central and South America.

American Alligator

Alligator mississippiensis
L 8–14'; W 300–1,000 lb **F**

appearance The alligator has a U-shaped snout; a dark brown, black, or greenish back; and a white belly.

food Young eat insects, larvae, and other small prey. Juveniles and adults move on to fish, birds, snakes, mammals, and any animal in or near water.

behavior Breeding begins in the spring. The female builds a nest in or near water to keep eggs warm during their 65-day incubation period.

range Southeastern U.S. coastal plain from Texas to Virginia, mostly in Florida, Louisiana; range is restricted to the U.S. Alligators inhabit a wide variety of wetlands, large and small. On occasion, they are found in suburban ponds, boat channels, and even swimming pools.

safety tip Alligators normally fear humans. But do not leave small children or pets alone near water where alligators might live.

Boy or Girl? As with many other reptiles and amphibians, the temperature of the eggs during incubation determines whether an alligator or crocodile will be male or female. If the temperature is below 86°F, the hatchlings will be mostly females. If it is higher than 93°, the hatchlings will be mostly males. Between those limits, the hatchlings will be about equally males and females. Unlike most reptiles, female alligators and crocodiles build their nests, guard the eggs, and then protect their hatchlings for three years. Male and female young look like tiny adults, but with yellow rings around the body.

Alligator hatchling breaks out of its egg.

food Crocodiles eat mostly fish but also birds, crabs, snails, frogs, small mammals, and occasionally carrion. Slow digestion allows a crocodile to go months without food in winter.

behavior Mating and female nest building take place during the winter dry season. Eggs incubate for 90 days and hatch at the start of the spring-summer wet season. Crocodiles mostly bask in the sun or quietly await their prey.

range Southernmost Florida in the United States; also Mexico, Central America, and South America.

Alligators and crocodiles are the only reptiles that walk with their legs under them.

[American Crocodile

Crocodylus acutus
L 8-14'; W 300-800 lb **S**

appearance The crocodile's V-shaped snout is longer and thinner than the alligator's; the body's brownish color is generally paler than that of the alligator. The young look like miniature adults. They lack the yellow bands that mark young alligators, but they usually show dark bands across the upper body that gradually fade.

did you know? American crocodiles are much less aggressive than their deservedly feared distant cousins in Africa and Australia? They are reclusive and are rarely seen except by scientists studying them.

Profile of an American crocodile

A turtle panorama shows different species of sea turtles.

Sea Turtles

After hatching from eggs on a beach, sea turtles crawl to the ocean. Males never return to land. Females return every two to four years to dig a hole in the sand, lay 100 to 200 eggs, bury them, and return to the sea. All sea turtle species are endangered.

Green Sea Turtle

Chelonia mydas
L 5'; W 700 lb **S**

appearance The top shell is colored in shades of brown, black, gray, and yellow; the bottom shell is yellowish white. This turtle's common name refers to the color of its body fat.

food These turtles eat sea grass and algae in shallow water; hatchlings eat tiny animals such as jellyfish, small shrimp, and crustaceans while they are swimming at sea.

behavior Breeding begins at age 25. Every two to four years females may migrate across oceans from feeding areas in shallow seas and estuaries to the breeding and nesting areas where the turtles themselves hatched. This species probably lives for more than 80 years. Adults swim up to 15 miles an hour and dive up to five minutes.

A green sea turtle, the largest hard-shelled species

range Mostly tropical areas of the Atlantic, Pacific, and Indian Oceans. Their normal preference is for regions of warm water where the temperatures do not fall below 68°F.

Hawksbill Sea Turtle

Eretmochelys imbricata
L 24"–45"; W 100–150 lb **S**

appearance The top shell is dark brown with yellow streaks and blotches; the bottom shell is yellow. The name refers to the turtle's hooked beak.

food The diet primarily consists of sponges (including species toxic to other animals), sea jellies, mollusks, fish, marine algae, and crustaceans.

behavior Females migrate long distances to lay their eggs on tropical beaches; mating occurs every two to three years.

range Primarily shorelines of tropical Atlantic, Pacific, Indian Oceans; U.S. nesting areas are on Puerto Rico, Virgin Islands, Hawaii's Big Island; a few nest in Florida.

Sea turtles dig nests and lay eggs at night.

Loggerhead Sea Turtle

Caretta caretta
L 3'; W 300 lb **S**

appearance The top shell is reddish brown and slightly heart shaped; the bottom shell is pale yellowish. The turtle's name refers to its massive head compared with its body.

food Loggerheads eat whelks, conchs, crabs, fish, and sometimes seaweed.

behavior This species can live for longer than 50 years. The female often returns thousands of miles to lay eggs on the beach where she hatched. The loggerhead can dive actively for 15 to 30 minutes, but when resting, it is able to remain underwater for as long as 4 hours.

range All but the most frigid waters in all oceans; often frequents coastal areas, estuaries, tidal rivers. Most common U.S. sea turtle species.

Hawksbill sea turtle

Snapping & Mud Turtles

Observing turtles is a good way to learn about reptiles, but you should not handle them or allow children to keep them as pets. Even healthy-looking turtles can carry dangerous salmonella bacteria. Unless you have expert advice, you should not handle any kind of wildlife because it could be disruptive to the animal.

An algae-blotched snapping turtle swims in a clear spring.

Eastern Snapping Turtle

Chelydra serpentina serpentina
L up to 19"; W 10–35 lb **F**

appearance This species has a flattened upper shell with knobby keels that smooth out with age, a small bottom shell, a large head, and a long tail with "sawteeth" on the upper side. Keels are ridges running from front to back on the upper shell.

food The eastern snapper is an important aquatic scavenger and active ambush hunter of anything it can swallow.

behavior Mating occurs from spring to fall. The female digs cavities on land to lay her eggs; she can store sperm for later fertilization, as do other reptiles. A snapper uses powerful jaws and a sharp "beak" to crush its prey.

range Shallow water in eastern half of North America, west to southern Alberta in the north and into central Texas in the south; closely related types in Central and South America.

safety tip

Wear shoes when wading in water where snappers might live, and keep your hands well away from their heads. A snapper can chop off a finger.

Maryland's diamondback terrapins were hunted to near extinction in the late 19th century? Their meat was in great demand for a gourmet stew. Snapping turtle meat is usually used in the U.S. today.

Diamondback Turtle/Terrapin

Malaclemys terrapin
L up to 12"; W 0.5–1.5 lb **F** **S**

appearance The top shell is black, brown, or grey; the bottom shell is yellow to green or black. The name refers to diamond shapes on the top shell.

food This turtle eats a variety of aquatic or marsh crabs, snails, mussels, and barnacles.

behavior Mating takes place during the summer. The female lays eggs on sandy beaches and hibernates during winter in cold locations.

range Northern subspecies near coast from Cape Cod to Cape Hatteras; southern subspecies near coast south to Florida Keys, west into southern Texas

Mississippi diamondback terrapin

Eastern Mud Turtle

Kinosternon subrubrum
L 3–4"; W 0.5 lb **F**

appearance The top shell is olive to dark brown; the bottom shell is yellow to brown. Males have a blunt spine at the end of the tail.

food The eastern mud turtle feeds on plants and animals, including earthworms, snails, insect larvae, dead fish, algae, and aquatic plants.

Eastern mud turtle

behavior Breeding occurs soon after the end of hibernation. The female lays eggs in a hole she has dug in sand or vegetation debris. Adults often wander far from water.

range Southwestern Connecticut and Long Island, south to southern Florida, west to central Texas and eastern Oklahoma, and north along the Mississippi Valley to southern Illinois and southwestern Indiana. Isolated population of the eastern subspecies in northwestern Indiana. Lives in shallow water in marshes, ponds, and other wetlands. It burrows into mud if the habitat dries up.

A pair of painted turtles bask on a log in the Pinery Provincial Park, Ontario, Canada.

Painted & Pond Turtles

How do turtles, tortoises, and terrapins differ? In North America, turtle can refer to any reptile that has a shell and lives in water. Some species spend more time on land than others. Tortoises live entirely on land. Terrapin, an Indian word, is used only for the diamondback terrapin. These words are used differently in other English-speaking nations.

Painted Turtle

Chrysemys picta
L 4–10"; W 10–17 oz **F**

appearance The top shell is smooth, and the body is brightly marked with various colors and patterns in different subspecies. Most have bright orange or red markings around the edges of the shell, and yellow stripes along the head, neck, and legs.

food This species is mainly carnivorous; it feeds on mollusks, worms, and minnows. An important freshwater scavenger, it helps to keep water clean. It eats more plants as it ages.

behavior The painted turtle hibernates for as long as October to March in the north. It may travel several miles overland in search of a better habitat or food. Often seen in groups basking

Juvenile eastern painted turtle

in the sun on rocks or logs at water's edge or above the water. Sometimes permits fairly close approach before escaping into the water.

range Common from coast to coast and from southern Canada to Gulf of Mexico and into Mexico. It can be found in marshes, ponds, lakes, and even in slow-moving rivers and streams with much vegetation.

An adult Pacific pond turtle underwater

Pacific Pond Turtle

Actinemys marmorata
L 6-8", W 1-2.4 lb **F**

appearance The upper shell is dark brown or dull olive, with or without darker streaks; the bottom shell is yellowish, sometimes with dark blotches. The upper shell of the adult is flat, broad, and smooth, with no keel (ridge) extending down the center. Head and legs are dull brownish, often blotched with dull yellowish spots.

food The diet varies widely and includes small animals such as fishes, tadpoles, and frogs, as well as plants such as algae and cattail roots; this turtle is also a scavenger, feeding on dead animals.

behavior This species reaches maturity at eight to ten years. It lives in marshes, swamps, rivers, ponds, and lakes. It favors habitats with rocks, logs, or aquatic vegetation suitable for basking in the sun. It may be aggressive to other turtles while it is basking, but when it is disturbed by a predator, it will quickly dive into the water.

range Limited to the West Coast of the United States and Mexico from western Washington to northern Baja California, but it is becoming rarer in the north. There is an isolated population in western Nevada.

did you **know** ?
the Pacific pond turtle is endangered in Washington and threatened in Oregon? Disease, habitat damage, and introduced predators such as bullfrogs and freshwater fish caused its decline.

A Pacific pond turtle hatchling emerges from its egg.

ON LOCATION

Everglades National Park, Florida

- Unique environment
- Best in winter
- Hundreds of animal species
- Hike, bike, and paddle tours

With 1.5 million acres, most of it wilderness, covering almost all of Florida's southern tip, Everglades National Park can be overwhelming. Where do you start planning a trip? The first decision is when to go, and the answer is easy: The only sensible time is between December and April. This is the dry season, and high and low temperatures average 77°F and 53°F. South Florida's rainy season— May through November—is also the park's mosquito season. Because the number of visitors declines at this time, the National Park Service reduces activities such as ranger-led tours and talks. But bicycle and canoe rentals are available to explore the park on your own.

A Unique Environment

The Everglades are not a swamp. They are really a 40- to 70-mile-wide river fed by rain and water from Lake Okeechobee and the Kissimmee River, which flows very slightly downhill at a rate of roughly 100 feet per day into Florida Bay. It flows through two distinct systems of low-lying areas called sloughs, which are deep enough to remain flooded for most of the year. The larger of these is Shark River Slough—called the River of Grass—on the west side. The smaller Taylor Slough is to the east.

The park includes the Western Hemisphere's largest area of mangrove swamps, which wind along the twisting rivers and creeks around the ocean coast. The brackish water in these waterways makes them nurseries for juveniles of many Gulf of Mexico and Florida Bay game fish. The sloughs, saltwater and freshwater marshes, and forested hammocks on slightly higher ground create a complex mix of ecosystems. These

Roseate spoonbills, snowy egrets, and white pelicans

Tourists take an airboat tour of the Everglades, a vast wilderness area.

environments are homes to 36 threatened or protected species, such as the Florida panther and the American crocodile, among the park's 350 species of birds, 300 species of freshwater and saltwater fish, 40 species of mammals, and 50 species of reptiles—including four species of venomous snakes that would rather leave you alone unless you surprise them. All of these species depend on a constant flow of fresh water. Unfortunately, people have been interfering with water flow into and out of the Everglades since the early 20th century. These changes have altered the park's ecosystems and threatened many of the park's inhabitants.

Meet the Park
Everglades National Park offers short hikes, bike rides, seven- or eight-day paddling trips through mangrove swamps, and camping on beaches that few people ever see except from an airplane. Park rangers lead short walking, bicycle, and canoe nature tours. The National Park Service has a list of permitted, licensed, and insured businesses offering longer walking, canoe, kayak, or pontoon boat tours ranging from a few hours to a few days and including backcountry camping. You will find fishing guides, trips into specific habitats, and nature photography instruction. The park prohibits airboat rides, but operators offer them in areas along U.S. Route 41 north of the park. A good way to get a taste of the park is to drive 38 miles from the main entrance at Homestead, south of Miami, to the Flamingo Visitor Center on Florida Bay. The center has several hiking and canoeing trails, a snack bar, a marina, and camping areas for recreational vehicles and tents. Houseboat, kayak, bicycle rentals, boat tours, and backcountry permits are available through the marina store.

Water Snakes

Only one species of North American water-oriented snake, the water moccasin—also called the cottonmouth—injects venom when it bites. Even though none of the other water-hunting snakes are venomous, you should not annoy them. Many are cranky and can inflict painful and potentially infectious bites. Water snakes belong to the Colubridae family, which has nearly 3,000 species worldwide (many of which do not live in water). North America's species range in size from eight-inch worm snakes to eight-foot black rat snakes.

Water moccasin

Water Moccasin

Agkistrodon piscivorus
L 20–48"; W up to 4 lb F S

appearance Juveniles have reddish brown bands that darken with age. Adults retain a hint of coloring or are uniformly black. The alternate name "cottonmouth" refers to the whitish lining of the mouth.

food Water moccasins eat mostly fish but also frogs, turtles, snakes, and birds when they are available. They kill prey by striking, biting, and injecting venom. Then they hold the prey in their coils until it is no longer struggling. Heat-sensing organs on its face enable this snake to detect warm-bodied prey even in the dark.

behavior This snake is solitary, stays within 0.3 mile of water, and hunts at night. Eggs develop within the female, and young are born alive. It is usually not aggressive, and will not attack unless it is agitated or surprised. But beware: The venom can be fatal.

range Swamps, lakes, rivers, and other watery habitats from the Florida Keys to the James River in Virginia, west into western Texas and the western edge of Missouri, and north as far as southern Illinois.

look for **water snakes sunning** on overhanging branches on cool days. They might drop into your boat and bite you, so always be vigilant.

Water Snakes

Nerodia species
L up to 4'; W up to 35 oz   F S

appearance The water snakes are several related species with heavy bodies. They are brown or olive green, with black, yellow, or cream markings. The belly of most species is yellow, white, gray, pink, or orange, and it may be either plain, neatly spotted, or blotched.

food Water snakes eat small, slow-moving fish, reptiles, frogs, tadpoles, worms, and small mammals that are commonly found at the water's edge or in the water.

behavior These snakes spend all or most of their time in or near water, and they are often found basking on tree branches over the water. Young are born alive. Some species are more

Water snake

 **did you know** ?

non-native populations of water snakes are established in some California rivers and lakes? Authorities suspect that pet owners released them because they are difficult to keep and care for.

active during the day, and others at night. Most water snakes are quite aggressive toward potential predators, including people, that come too close.

range At least seven species and many subspecies occur in overlapping ranges from southern Canada to the Gulf of Mexico and into Mexico, and west as far as Kansas and Nebraska. Their diverse habitats include well-vegetated marshes, lakes, streams, rivers, ponds, canals, ditches, bayous, and estuaries.

Water snake

Mud Puppies & Salamanders

Despite their appearance, mud puppies, salamanders, and newts are not lizards. They are amphibians, which live both on land and in water, and they are related to frogs and toads. Lizards live strictly on land, and many require little or no water except what they get from their food.

Mud Puppy

Necturus maculosus
L 8–19" **F**

appearance The mud puppy has bushy, dark red external gills that wave in the current, a flat head, stubby legs, and a gray or brownish gray body with dark spots. A long tail fin is sometimes orange or reddish.

food Mud puppies feed on small fish, crayfish, aquatic insects, worms, and snails—almost anything they are able to catch in the water.

behavior This animal is a bottom dweller that hunts at night. The female lays 50 to 100 eggs and guards them until they hatch. The mud puppy is an entirely aquatic salamander with gills.

range Southern Canada to Georgia, inlands of East Coast states west into the Mississippi Valley. Mud puppy habitats vary greatly; some populations are adapted to cold, clear waters and others to warm, muddy waters.

listen for **sounds that resemble** the yelps of a puppy. These are made by mud puppies, which some people call waterdogs.

Coastal (Pacific) Giant Salamander

Dicamptodon tenebrosus
L up to 13" **F**

appearance The giant salamander has a large body with a massive head, stout limbs, and a flat tail. Its body is dark brown to nearly black, with irregular light brown spotting or mottling. The skin is smooth.

food This species eats anything it can overpower and fit into its mouth, including large insects, mice, smaller salamanders, and small snakes. Larvae may eat tadpoles and small insects.

behavior The female lays eggs in slow-moving streams and guards them until they hatch. Adults may bite, and their skin exudes harmful secretions.

Mud puppy

A coastal giant salamander rests on a fallen tree.

range Northern California into British Columbia and east to the Cascade Mountains. The habitats include rivers, streams, and humid forests that are not far from water.

Eastern Newt

Notophthalmus viridescens
L up to 5" ■

appearance Larvae are dark green, juveniles are bright orange-red, and adults are olive green with red spots. The species is alternately named the red-spotted newt.

food Juveniles eat small invertebrates; adults eat insects, worms, small crustaceans, and mollusks.

behavior When it is approximately a year and a half old, this newt loses its gills and grows paddle-shaped hind legs. This species forages in shallow water or on the forest floor.

range Eastern U.S. (except southern Florida) west to Minnesota, eastern Kansas and Texas.

Hellbender

Cryptobranchus alleganiensis
L up to 30"; W up to 5.5 lb ■

appearance The hellbender is the largest North American salamander. Adults are greenish, yellowish brown, or gray with black spots or blotches.

food The diet includes crayfish, small fish, mollusks, and worms, including some with slime that is normally noxious to many predators.

Hellbender

behavior The hellbender lives in fast-moving rivers and large streams under rocks or snags. It forms breeding aggregations in and around nest sites, such as burrows in streamside banks or under rocks. The male protects eggs from being eaten by other hellbenders.

range Appalachians and adjacent areas from southern New York into northern Alabama.

safety tip **Handling salamanders** can be dangerous. Some species contain a poison, tetrodotoxin, which can irritate your skin.

Eastern newt

Frogs

Spring peeper

Frogs are a delightful feature of almost every clean-water pond in North America. Observing egg masses, tadpoles, and adult frogs is a good introduction to amphibians—and to nature at large. Frogs are fun to hear too. From the low thrum of a bullfrog to the high-pitched ring of a spring peeper, each species has a distinctive song, which males deliver nonstop as they advertise for mates in the spring.

Bullfrog

Bullfrog

Rana catesbeiana
L 3.5-6" **F**

appearance The bullfrog has smooth skin and long hind legs. It is greenish, brownish, or blackish above; its belly is white to yellow. With large, bulging eyes, it can see well both in the day-time and nocturnally.

food Bullfrogs eat any animal they can overpower and swallow—rodents, turtles, smaller frogs, small birds, insects, fish, tadpoles, snails, young snakes, crayfish, and beetles. They quickly swallow their prey whole.

behavior The bullfrog breeds in vegetation-choked shallows, and it is usually found during the day at the water's edge. Tadpoles may become adult frogs in their first year or may overwinter and mature in their second year. Its twangy *jug'o-rum* vocalization is famous.

range U.S. from Atlantic coast west into western Mississippi Valley, along West Coast, and parts of the desert in the Southwest and Great Basin.

Spring Peeper

Pseudacris crucifer
L 0.75-1.25" **F**

appearance This small frog is yellow, brown, gray, or olive with a darker X mark on its back, which may not be prominent.

food The diet consists of arachnids such as spiders. It also eats ants, beetles, and other small, slow-moving invertebrates.

behavior Spring peepers live in wetlands at sources of streams and ponds, and along small streams.

A spring peeper sits in a calla lily.

listen for **a chorus** of high-pitched whistles, which are a sign of spring as peepers emerge from hibernation and advertise for mates.

range Widespread east of a line from eastern Texas north to eastern Manitoba, and across southeastern Canada.

Northern Leopard Frog

Rana pipiens
L 3–5" **F**

appearance This frog is greenish brown with an array of irregular dark spots on its back and legs; it is one of 14 similar species with similar markings.

food This species is an ambush hunter that eats almost anything that fits in its mouth, including beetles, crickets, flies, worms, and smaller frogs.

A leopard frog in garden vegetation

behavior The northern leopard frog lives in ponds, swamps, slow-moving streams, and open fields; it is sometimes called a meadow frog.

range Various species live in most of temperate and subtropical North America; pollution is likely a cause of rapid decline in many areas.

Life Cycle of a Frog

The change from tadpoles, which look like tiny fish, into adult frogs is easy to observe during regular trips to the same pond, or even at home if you make sure to create the right conditions. The period from hatching to adult is approximately 16 weeks.

Eggs hatch 4–9 weeks: begins to eat insects, head grows larger, legs emerge

12–16 weeks: loses last of tail, becomes adult

9–12 weeks: begins to lose tail, moves onto land

Toads

Adult toads live farther from the water's edge than other amphibians do; they return to the water only for breeding. Land-living toads can be a gardener's best friends by gobbling up pests such as cutworms, beetles, and other plant nibblers. In fact, many gardeners encourage toads to take up residence by making or buying toad houses to provide cool daytime shelter for toads that come out at night to seek their prey.

American toad

American Toad

Bufo americanus
L 2–4.5" **F**

appearance This common toad of gardens and woodlands has a tan, brown, reddish, or olive green back, with brownish spots, pale areas, and warts that range from brownish to reddish orange. A pale stripe usually runs down the center of the back. The belly is white to yellow, usually with black or gray spots.

food The diet includes insects, crickets, mealworms, earthworms, ants, spiders, flies, slugs, centipedes, moths, and other small invertebrates.

behavior This species is solitary except in the spring breeding season, when males go to shallow water to call out to females. Tadpoles hatch in 3 to 12 days, become adults in 50 to 65 days, and move away from water. They burrow into the ground to hibernate.

range At least three subspecies across Canada from southeastern Manitoba to Labrador and southward to Oklahoma in the West and to Louisiana, Mississippi, and Georgia in the East—many diverse climatic regions.

did you know ? **warts are caused** by the human papillomavirus, not by touching a toad? A toad's "warts" are bumps, part of its camouflage, but some of them are glands that excrete a poison that can irritate your hands.

American toads lay eggs at the river's edge.

Frogs vs. Toads

When you see a frog or a toad, your first question may be, "Which is it?"

In general, frogs have smooth, clammy skin, and toads have drier, bumpy skin. Toads, which usually have shorter legs compared with frogs, tend to walk or hop. Frogs' long rear legs help make them phenomenal jumpers, and they are better at escaping predators.

All native North American toads and some frogs secrete mild toxins that can be irritating but are not fatal, even to animals such as dogs and cats. Still, to avoid irritation, it is a good idea to wash your hands after handling any toad or frog before touching your eyes or mouth.

The non-native giant toad (up to 6" long), which is native to Central and South America, has been imported to eat sugarcane pests in southern Florida, Hawaii, and Puerto Rico. It has toxic glands that can kill cats and dogs that bite it. It is a threat to native frogs and toads because it is big enough to eat them—and does.

Frogs' long hind legs make them powerful jumpers.

A western toad sitting in grass

Western Toad

Anaxyrus boreas
L 2–5" F

appearance The western toad is dusky gray or greenish with darker blotches and a white or cream stripe down the middle of the back.

food This species eats bees, beetles, ants, spiders, crayfish, sow bugs, grasshoppers, butterflies, and flies.

behavior Adults live mostly on land. Their hibernation and active periods depend on elevation and latitude. They hunt mostly at night at low elevations but during the day at high elevations. This species may excavate its own burrow or use a small rodent's old burrow.

range Western North America from southern Alaska across British Columbia, southward through Washington and Oregon into central California, and eastward from the Pacific coast to the Rocky Mountains; also along southern California coast south into Baja California, Mexico.

Tree Frogs

A tree frog is any species that has evolved to spend most of its adult life hunting in trees. Tree frogs are found around the world. The species are not all related to one another, though they share several traits. All are small enough to climb on tiny branches, and all have sticky pads at the ends of their fingers and toes. Many change color for camouflage.

A young boy studies a Pacific tree frog. These frogs use sticky pads on their toes for climbing.

Pacific Tree Frog

Pseudacris regilla
L up to 2" **F**

appearance Most are green or brown, but they can be other colors; this species is identified by its black or dark brown stripe from the nose across the eye to the shoulder. Head and eyes are relatively large, the skin is smooth and moist, and the legs are long and slim.

food Diet includes spiders, beetles, flies, insects, and other arthropods; catches insects with a sticky tongue.

behavior The Pacific tree frog can climb plants and other surfaces in order to hunt. It is mostly nocturnal, and it usually hides and is difficult to spot when not moving.

range Pacific states from northern California to Canada; in diverse moist and dry habitats from sea level up to elevations as high as 10,000 feet in mountains.

listen for **tiny tree frogs** producing loud mating calls. These are produced by sound sacs on the neck, which can expand to three times the size of the head.

A green tree frog on foliage in a marshy area

Green Tree Frog

Hyla cinerea
L up to 2.5" **F**

appearance Usually a shade of green, this frog may have small patches of yellow or white; it is distinguished by a light-colored stripe from the upper lip along the sides to the groin.

food Green tree frogs eat mostly crickets but also mealworms, wax worms, fruit flies, fireflies, pinky mice, and some fruit.

behavior This species prefers ponds, lakes, and streams with plentiful floating vegetation in forests with open canopies; mating occurs from April to August. Hundreds form choruses giving vocalizations that have been compared to a the sound of a cowbell.

range One of most common tree frogs in the southeastern U.S. from eastern Texas north to Illinois, and Atlantic coast north to Maryland.

Gray Tree Frog

Hyla versicolor
L up to 2" **F**

appearance The color can vary from green to light green-gray, gray, brown, or dark brown. There is a large white spot below each eye. This species can change color from nearly white to nearly black for purposes of camouflage against its surroundings.

food This frog eats most types of insects, mites, spiders, plant lice, and snails. Adults are nocturnal hunters. Tadpoles graze on algae and detritus in a pond.

behavior A fine climber, it can use its toe pads to scale glass. It hunts in forests and breeds from May to July.

range Most of the eastern half of the United States, in the north from Manitoba to Maine, south to western Texas and northern Florida.

A gray tree frog blends in to the bark of a tree.

ON LOCATION

Great Smoky Mountains National Park

- World salamander capital
- Most visited national park
- Stunning mountain views
- Lush temperate forests

A good way to experience the Great Smoky Mountains is to take time absorbing both the big picture and the close-up views. Hikers can savor the big picture by viewing the sunset from a comfortable chair after a hardy dinner at the hikers-only LeConte Lodge atop 6,593-foot Mount LeConte. If you would rather forego a five-hour hike up a mountain, drive to the top of 6,643-foot Clingmans Dome, the park's highest mountain, to take in the view. For close-ups you could search for salamanders in the park's cool streams. The Smoky Mountains are called the Salamander Capital of the World.

Salamanders and the Smokies

The National Park Service says that on any given day, the great majority of vertebrates in the park will

Great Smoky Mountains National Park, in Tennessee, is known as the Salamander Capital of the World.

be salamanders. This encompasses other amphibians, reptiles, birds, and mammals—including you and other visitors. Beyond the unusual number of salamanders, more than 17,000 known species of plants and animals live in the park, and many remain to be documented. The unbroken range of 5,000-feet-plus mountains along the park's spine, which forms the Tennessee-North Carolina border, sets the stage by forcing warm, humid air from over the Atlantic Ocean or the Gulf of Mexico to rise and cool. Cooling condenses some of the moist air into rain or snow that falls on the mountains. Each year 55 to 85 inches of precipitation (rain and melted snow and ice) falls on the park. Unlike almost all other Appalachian locations, loggers never stripped the trees from a quarter of today's park. These virtually untouched areas continue to harbor a wide variety of life-forms.

Park elevations range from 875 to 6,643 feet. If you hiked from the lowest to the highest, you would pass through the same variety of forest environments that you would see if you traveled from Georgia to Maine. The mountains' steep, jumbled topography and wetness create extreme variations in temperature and moisture that add up to a variety of habitats, most of which benefit from soils rich in minerals. These habitats support multitudes of invertebrates that salamanders eat. Another source of the park's diversity is plants and animals that retreated from advancing ice during the last ice age. They found suitable habitats in the higher parts of the Smokies and stayed after the glaciers retreated. Over time new species evolved to fit into particular environmental niches.

A northern red salamander in the Great Smokies

Finding Salamanders

The nickname Salamander Capital of the World might make you expect to see these creatures crawling around all over the park. You will not. Salamanders are rather reclusive, and they tend to hang out under rocks or logs in wet areas near streams and places where water seeps from the ground. You could start searching for them by reviewing information at the park's visitor centers or purchasing a book on the park's amphibians. The National Park Service discourages visitors from turning over logs and rocks in the park, but you could go on a ranger-led tour that includes learning about salamanders.

If you plan to travel to the park in late April, check out the annual Wildflower Pilgrimage, which usually includes several salamander identification trips as well as hikes to enjoy blooming wildflowers. Visiting in the spring helps you avoid the larger summer crowds: Great Smoky Mountains National Park is the country's most visited national park. Also look for the family field trips run by the University of Tennessee's Smoky Mountain Field School, such as the trips called Sensational Salamanders or Amphibians and Reptiles of the Smokies.

Permit me to place you on the Mississippi, on which you may float gently along, while approaching winter brings millions of water-fowl on whistling wings, from the countries of the north, to seek a milder climate in which to sojourn for a season.

—JOHN JAMES AUDUBON AND JAMES MACGILLIVRAY,
ORNITHOLOGICAL BIOGRAPHY

[6]

BIRDS

Watchful, a matched pair of bald eagles dip
their talons in seawater on a beach at low tide.

Geese & Swans

Geese and swans symbolized wildlife conservation a century ago, but today some waterfowl are considered pests. Booming populations of nonmigratory, feral Canada geese deposit waste on grass and in waterways, and snow geese threaten delicate tundra ecosystems. Mute swans are destroying wetland habitats by uprooting emergent vegetation.

Canada Goose

Branta canadensis
L 30–43"; W 5–12 lb **F** **S**

appearance This honker is the most common and familiar goose in North America. It is large but variable in size. It has a black head and neck with a white chinstrap; otherwise it is mostly gray-brown and whitish below.

food Canada geese are often seen grazing on lawns and other grassy areas, but they also feed on aquatic plants. Large winter flocks are attracted to agricultural fields with waste grain.

behavior Pairs may stay together for life. Goslings fly at about nine weeks, and parents and young remain together until the following spring.

Canada goose

These geese migrate in V-formation flocks, and they call to each other while in flight.

range Widespread, common species and year-round resident; most birds seen in parks, reservoirs, and golf courses are nonmigratory and feral; migratory birds breed mainly in Canada and spend the winter months in the lower 48.

Snow Goose

Chen caerulescens
L 26–33"; W 5.5–7.5 lb **F** **S**

appearance There are two color morphs. The white morph is all white except for black primaries (outer wings); the less common blue morph has a white head and neck, a dark brown body, and gray and black wings.

food Snow geese eat tundra vegetation and some insects when breeding in the far north; they feed on aquatic vegetation and waste grain in winter.

behavior This species forms a lifelong pair bond. It migrates in high-flying, noisy flocks. In winter it forms large flocks, sometimes with other species of geese.

range Breeds from May to August mainly in Canadian Arctic; winters in warm parts of North America from southwestern British Columbia through parts of the United States to Mexico.

Snow geese in flight

Choosing Binoculars

Binoculars greatly increase your ability to observe wildlife, especially birds. They should be light enough for you to hold them up comfortably for extended periods of time. Numbers on binoculars, such as 7x35 or 10x50, refer to the magnifying power (first number) and the diameter of the outer lens in millimeters (the larger the number, the brighter the image). Most people feel comfortable with 7x or 8x magnification with an objective lens of 35 to 45 millimeters. Choose a pair with a single, central focus wheel that feels natural to turn. The prisms should be made of BaK-4 glass and have multiple lens coatings to transmit light well. If possible, try out the binoculars before making a purchase.

Waterproof binoculars are a plus.

Mute Swan

Cygnus olor
L 60"; W to 26 lb **F** **S**

appearance This is a very large and graceful swan with all-white plumage and a black knob at the base of its orange bill. It holds its long neck in an S-shaped curve.

food The diet consists of aquatic vegetation, usually pulled up from the bottom (its long neck allows it to reach deeper vegetation than other waterfowl); it also eats grain and grasses on land.

behavior The mute swan builds a large nest of vegetation in shallow water. It is an aggressive defender of its nest and young. This species can greatly alter a wetland's ecological balance by uprooting vegetation and is considered a pest in many areas.

range European species introduced in U.S. and often seen in parks, increasing along the East Coast with feral populations from Cape Cod to Virginia, also around the Great Lakes and on Vancouver Island, British Columbia.

A mute swan with cygnets

Ducks

At first glance, all ducks look very similar. Nevertheless, different kinds live in dissimilar ways. Perching ducks have sharp claws that enable them to perch in trees. Dabbling ducks—puddle ducks—feed by tipping their tails up to grab food under the water. Mergansers dive underwater and use their long, serrated bills to catch fish and other aquatic animals.

The male wood duck is one of the most colorful of North American waterfowl.

Wood Duck

Aix sponsa
L 18.5"; W 1.3 lb **F**

appearance The male has glossy, colorful plumage with an intricate design and a sleek crest; the less ornate female has a short crest and a white teardrop around the eye. In late summer, males molt into eclipse plumage and look very similar to females.

food The diet mainly consists of berries, acorns, seeds, and insects. Wood ducks are usually seen foraging in pairs or family groups. Young birds leave the nest one day after hatching and can feed themselves.

behavior This species nests in or near wooded swamps, shallow lakes, marshes, or ponds, usually in tree cavities or nest boxes. Pairs are often detected by the loud squeal of the female as they take flight; males are less vocal.

range Year-round resident in the South and on the West Coast; birds that breed farther north migrate south in the fall.

look for **wood duck nest boxes** located on poles in shallow water. Nest boxes helped this species rebound from low numbers in the early 1900s.

did you know ? **colloquial names for ducks** were once common? Some colorful alternative names for the hooded merganser were cock-robin duck, frog-duck, little spikebill, mosshead, and pond fisher, among many others.

A male mallard in flight

Mallard

Anas platyrhynchos
L 23"; W 2.4 lb **F** **S**

appearance The male has a metallic green head, a white neck ring, and a yellow bill. The female is very different, with a mottled, brownish body and an orange and black bill.

food Mallards eat aquatic weeds, grasses, and seeds. It sometimes feeds on insects, fish, earthworms, frogs, tadpoles, and snails. It is often seen with its tail in the air and its head underwater; this behavior is known as tipping up.

behavior A dabbling duck, the mallard nests in areas safe from predators, sometimes in such unusual locations as roof gardens or flower boxes on window ledges.

range Makes use of almost any size body of water from Alaska to Mexico; birds breeding in most of Alaska and Canada migrate to the lower 48 in fall.

Hooded Merganser

Lophodytes cucullatus
L 18"; W 1.3 lb **F**

appearance This is a small diving duck. The adult male has a black crest at the back of the head, with a white center and a black back; the female has a reddish crest but is otherwise brownish overall, with a paler breast. The male's crest can be fully expanded, showing a large white crescent, or it can be folded back, showing just a stripe of white.

food This species eats mostly fish, pursued underwater, and some aquatic invertebrates, such as insect larvae, snails, and small crustaceans.

behavior A hooded merganser's nest is usually located in a tree cavity near water. The female incubates and raises young alone; chicks leave the nest with their mother within a day of hatching. This duck is generally silent, but it makes a froglike growl during courtship displays.

range Common in eastern and central North America, uncommon in West; northern birds migrate south in fall.

The male hooded merganser has a large crest.

Common loons emit eerie wailing yodels during breeding season.

Loons, Grebes & Coots

When exploring ponds or lakes during the breeding season, look for loons, grebes, and coots diving for food. You might even want to time how long one stays underwater; the average loon dive lasts about 45 seconds. Loons' and grebes' legs are set so far back that they cannot walk normally.

Common Loon

Gavia immer
L 32"; W 9 lb **F** **S**

appearance A large diving bird with a stout, spiky bill, the common loon is almost always seen swimming. In the summer, adults have a blackish green head and neck and a bold black-and-white back pattern; winter plumage is blackish brown with a white throat and a pale bill.

food Loons mostly eat fish pursued underwater. They can dive to almost 200 feet but usually pursue their prey much closer to the surface. A specially adapted bill helps grasp prey.

behavior This species nests in vegetation near the water's edge, often on an island in an undisturbed lake. Parents swim with their young until they can fly at about two months.

range Breeds in Alaska, Canada, and the northern tier of states; in winter, found in coastal areas of the East, South, and West, and large inland water bodies.

listen for
loon calls, signature sounds of the north. There are four distinct calls: a hoot, a call resembling insane laughter, a yodel, and a wail.

look for aggressive, territorial male coots fighting, with one bird trying to hold his opponent underwater. The vanquished male usually flees with a splashy takeoff.

Pied-Billed Grebe

Podilymbus podiceps
L 13.5"; W 1 lb **F S**

appearance This chunky, ducklike bird has a thick, chickenlike bill; in the summer, a black band encircles the bill.

food This species dives for aquatic invertebrates as well as small fish, tadpoles, and frogs.

behavior Rarely seen in flight, the pied-billed grebe builds a bowl-like nest on floating vegetation and covers the eggs with vegetation when leaving the nest. The downy, striped young sometimes ride on their parents' backs.

range Common breeder across North America on lakes and ponds with emergent vegetation; in winter migrates to warmer regions, sometimes on saltwater bays and harbors.

An American coot adult with chicks

American Coot

Fulica americana
L 16"; W 1.4 lb **F S**

appearance The American coot is a chunky, ducklike bird with dark gray plumage and a white, chickenlike bill. Coots are not related to ducks (or chickens), but they are related to secretive marsh birds called rails. The coot's feet are not webbed, but its toes have expanded lobes to aid in paddling.

food This species dives for food and also forages on land and mudflats. It eats plants, arthropods, fish, and other aquatic animals.

behavior The bird is seen in wetlands, mudflats, brackish marshes, and often on golf courses; it can occur in large flocks. Coots expend lots of energy to become airborne, pedaling frantically across the water with their feet before lifting off. Once in the air, their flight is strong and fast. Coots make a variety of grunting and cackling notes, and they are vocal day or night.

range Abundant, year-round resident in much of the U.S.; northern interior birds move to coasts and southern U.S. in winter.

A pied-billed grebe with its young

Pelicans & Cormorants

Pelicans and cormorants have evolved different ways of catching fish. Brown pelicans fly over the water in lines and then plunge down for a meal. White pelicans, often in organized groups, herd fish as they swim and then scoop up the fish with their bills. Cormorants chase fish underwater and capture them with the help of a strongly hooked bill.

A brown pelican in flight

Brown Pelican

Pelecanus occidentalis
L 48"; W 8.5 lb **S**

appearance The brown pelican is a very large waterbird with a grayish brown body, a black belly, a white or yellowish head, and a long bill with a skin pouch. Juveniles are brownish with a dark head and a pale belly.

food Pelicans eat fish. They make spectacular aerial dives to catch underwater prey. They dive from as high as 50 feet and scoop up as much as 2.5 gallons of water. Upon surfacing, water drains from the pouch, and the bird swallows the fish with a backward flip of its head.

behavior This species prefers nesting on islands. It spends a lot of time loafing on piers, barges, rocky islets, and mangroves, often in large groups.

range Resident along Gulf Coast and on Atlantic coast north to about Maryland; also breeds off southern California and wanders north along the Pacific coast after breeding.

White Pelican

Pelecanus erythrorhynchos
L 62"; W 15–20 lb **F S**

appearance This huge waterbird has bright white plumage and black wingtips. Its long, yellow to orange bill has a classic pelican shape. Breeding individuals grow a curious, hornlike knob on the top of their bills, about two-thirds of the way to the tip. Young birds are similar to adults, but their wings have brown streaks.

food The white pelican eats fish, crayfish, and other small aquatic animals.

White pelican

behavior This species breeds in large colonies on inland lakes, and it feeds in cooperative groups. It sometimes soars high overhead, especially when migrating.

range Breeds throughout the inland West and prairie regions, with the largest colonies in Canada; winters on coastal bays and large inland lakes mostly in California, Texas, the Gulf Coast, and Florida.

A double-crested cormorant dries its wings.

Double-crested Cormorant

Phalacrocorax auritus
L 32", W 3–5 lb

appearance The adult is all black. Juveniles are gray-brown with paler underparts that darken over the first two years of their lives. All ages have yellowish facial skin and throats, as well as blackish legs with fully webbed toes. Breeding birds have two small crests located on either side of the head.

food The diet mainly consists of fish caught underwater. The bird uses its webbed feet for underwater propulsion; its snakelike neck and strong hooked bill assist in capturing slippery prey.

behavior This cormorant breeds in coastal areas and near inland rivers and lakes. It builds stick nests in trees, in cliff edges, or on the ground. It is often seen perched with its wings held out to dry.

range Breeds across North America from coastal Alaska to Florida and Mexico; winters in ice-free locations on coasts and inland bodies of water.

Bill Shapes When trying to identify a bird, its physical attributes, such as color, shape, and size, are important field marks. The shape of a bird's bill sometimes not only helps you identify it, but also tells you how it makes its living. This realization helped set Charles Darwin's thoughts on the road to developing the theory of evolution by natural selection, as he contemplated the varieties of finches' bills and how they were used in the Galápagos Islands in 1835. The birds illustrated below are sandpipers with bills of differing lengths and curvatures, which allow them to feed on different prey items.

Spotted sandpiper

Wilson's snipe

Whimbrel

Herons & Egrets

Most herons and egrets can be readily identified by a combination of plumage coloration, overall size, and details of the head and bill. But they sometimes require a close look. For instance, the differences between great and snowy egrets are not dramatic. All herons and egrets fly with their necks held in an S curve. This characteristic separates them from cranes, which fly with their necks held straight out.

Great Blue Heron

Ardea herodias
L 46"; W 5.5 lb F S

appearance This large heron has a slate-gray body with chestnut and black accents, a very long neck, and dark legs. An all-white morph with dull yellow legs occurs in southern Florida, but it is uncommon even there.

food The diet includes small fish, crabs, snakes, and small mammals. The bird finds prey by walking slowly or standing still and waiting. Its long, kinked neck allows it to make lightning-fast strikes with its daggerlike bill.

behavior The great blue heron nests in colonies, often with other species. It constructs a stick nest in a tree or large shrub.

range Common, year-round resident in much of the U.S.; frozen water forces northernmost breeders to move south or to coastal areas in winter.

A great egret on the wing

Great blue heron

Great Egret

Ardea alba
L 39"; W 2 lb F S

appearance This large all-white egret has black legs and feet and a yellow bill. Breeding birds have lacy plumes (known as aigrettes) extending from the back past the end of the tail.

food The great egret eats fish, frogs, small mammals, and occasionally small reptiles and insects. Its long legs allow it to wade in deeper water than the similar-looking snowy egret.

behavior This species is often seen alone, but in favorable locations small flocks occur. It sometimes hunts in fields and pastures. It nests colonially in trees or shrubs, often with other wading birds.

range Year-round resident in coastal areas and in the South. Northern and interior breeders migrate to ice-free areas for the winter months.

Snowy Egret
Egretta thula
L 24"; W 13 oz F S

appearance This all-white egret, which is smaller than the great egret, has a black bill and legs and yellow feet. Breeding birds have elegant, wispy plumes on their crowns, necks, and backs.

food The diet includes fish, crustaceans, insects, and small reptiles. The bird stalks prey in shallow water, and it is often seen patting the water's surface with its foot to stir up fish and other prey, which it captures with a rapid strike.

behavior Through the early 1900s this bird's breeding plumes were used in the millinery trade, resulting in the near decimation of the species.

Snowy egret

range Common in most coastal U.S. states; also breeds in the interior West and in much of the South. Northern and most interior breeders move coastward for the winter.

The green heron has short orange legs.

Green Heron
Butorides virescens
L 18"; W 7 oz F S

appearance The green heron is small and dark with short, yellow-orange legs. Adult plumage is greenish gray above with a blackish cap and a chestnut neck; the juvenile is plainer and brownish, with heavily streaked underparts

food This bird eats small fish, frogs, and aquatic arthropods. One of the few North American birds to use tools, it places a lure, such as a leaf or small twig, on the water and waits patiently for a fish to investigate—a fatal mistake for the fish.

behavior Often quiet and inconspicuous, green herons are most active at dusk and dawn; when flushed, they give a loud, sharp *skeow* call.

range Common and widespread breeder in much of U.S., but mostly absent from the interior West.

An osprey takes off from the water with a trout in its talons.

Raptors

Raptors—ospreys, eagles, and hawks—often soar on rising air currents to stay aloft with little effort and use their keen vision to spot prey below. To catch their dinner by surprise, ospreys plunge into the water, while bald eagles usually swoop down in a low glide—or pirate a meal from an osprey. Both raptors have large and extremely sharp talons to hold their favorite prey: a slippery fish.

Osprey

Pandion haliaetus
L 22-25"; W 3.5 lb **F** **S**

appearance The osprey is dark brown above and white below, with a white head and a prominent dark eye stripe. In flight, its large wings, bent back at the wrist, present a distinctive crook-winged silhouette.

food Known to many as the fish hawk, the osprey is almost exclusively a fish eater. It dives talons-first into the water, makes a large splash, and sometimes completely submerges. It is capable of capturing a large fish and then carrying its prey to a favorite perch or a nest with hungry young, where the fish is torn apart and consumed.

behavior Invariably nesting near a body of fresh water or salt water, this raptor builds a large stick-and-sod nest in a tree, on a utility pole, or on a channel marker. A man-made nest platform

> **did you know?**
> **an osprey's eyes** are adapted to seeing fish under the water from as high as 130 feet in the air? It then plunges talons-first into the water to grab the fish and lift it into the air.

A bald eagle, the national bird of the United States

is another favorite location and is often used for many nesting seasons.

range Breeds from Alaska across northern Canada to the Atlantic, around the Great Lakes, in the northwest and northern Rockies in U.S., along Atlantic coast to Florida. Most North American breeders migrate to Central and South America for the winter.

Bald Eagle

Haliaeetus leucocephalus
l 31-37"; W 9.5 lb **F C**

appearance The official symbol of the United States, the adult is readily identified by its white head and tail, dark brown body, and large yellow bill. The juvenile is quite different; it is mostly dark with white streaking on the underwings and at the base of the tail. The bald eagle takes four years to reach adult plumage. The sexes have identical plumage, but the female is larger than the male.

food This eagle feeds primarily on fish in the breeding season, but it also captures ducks, geese, and other large birds. It also feeds on carrion and road kill, especially in the winter.

behavior Except during migration, this species is always seen near water,

> **look for** **a bald eagle chasing** an osprey with a fish and taking its prey, a behavior known as kleptoparasitism. The bald eagle is a bit of a bully in the bird world.

primarily seacoasts, large lakes, or rivers. It builds a large nest—typically five feet in diameter—in a tall tree or on a cliff. Its call is a weak, high-pitched whistle.

range Occurs across North America. Northern breeders move to the coast or south of the freeze line to find open water—the outflow of a hydroelectric dam is a favorite location.

A juvenile bald eagle on a beach with its prey

ON LOCATION

Fraser River, British Columbia, Canada

- Largest West Coast estuary
- Millions of migrating birds
- From waterfalls to mudflats
- Backcountry exploration

A couple frolics under a Fraser River waterfall.

Canada's Fraser River squeezes a complete range of river environments, from high mountains to a saltwater estuary, into a relatively short distance that allows visitors to explore an entire river during one vacation. The 851-mile-long river begins in Fraser Pass on the western slopes of the Rocky Mountains in the southeast corner of Mount Robson Provincial Park. It meanders first toward the north to the outskirts of Prince George, British Columbia, where it turns to the south. On its way south the river cuts in to the Fraser Plateau and forms the 3,300-feet-deep Fraser Canyon before turning to the west to reach the Strait of Georgia at Vancouver, Canada's third largest metropolitan area behind Toronto and Montreal. The Frasier River Estuary, where the river mixes with salt water, is the largest estuary on North America's Pacific coast and one of the world's biologically richest estuaries.

Nature on a Big City's Edge

Each year the Fraser River picks up an estimated 20 million tons of sand and carries it to the Straits of Georgia. Over thousands of years this sediment

Sunrise near Steveston Village, a historic salmon-canning center at the Fraser River's mouth

has built up a delta between central Vancouver and the U.S. border to the south. The delta includes Vancouver's southern suburbs, farmland, and wildlife preserves, and it has 116 square miles of adjoining intertidal marshes and mudflats. Several million waterfowl and shorebirds from Arctic and sub Arctic breeding grounds in Siberia, Alaska, and the Yukon stop at the Fraser Estuary on the way to California, Mexico, Central and South America, or the South Pacific. Many feed on the Roberts Bank mudflats in the middle of the estuary harbor, which are several miles wide at low tide and offer the birds a rich menu of invertebrates. Other birds forage nearby cultivated fields for terrestrial insects and worms. You will also see numerous raptors, including red-tailed hawks, peregrine falcons, and bald eagles. Good viewing areas are the adjoining Alaksen National Wildlife Area; the George C. Reifel Migratory Bird Sanctuary, about a 20-mile drive

from the center of Vancouver; and the South Arm Marshes Wildlife Management Area a few miles further south.

Upstream Explorations

You could explore inland Fraser River areas by driving roughly 70 miles west and north from Vancouver on the Trans-Canada Highway to Hell's Gate in Fraser Canyon. Here the canyon's walls are 3,300 feet high, and the Trans-Canada Highway and rail lines run through it. You can enjoy this part of the Fraser no matter how much outdoor experience you have. For instance if you are there in October, you could take an aerial tram to the canyon's bottom to watch salmon climb fish ladders. If you are comfortable in wilderness on unmarked trails, the area's provincial parks offer extended backcountry hikes or shorter explorations, such as hiking along Mehatl Creek to Mehatl Falls, where chinook salmon, bull trout, and rainbow trout spawn in the creek below the falls.

Shorebirds

Shorebirds are the easiest to observe among water's-edge wildlife. They patrol beaches, mudflats, or shallow water and find food in a variety of adjoining environmental niches. Some use sensitive nerves in their bills to find buried prey. Various bill lengths and shapes allow others to extract creatures that burrow deep. Yet other shorebirds pull prey up from closer to the water's surface. Straight, strong bills allow some to eat crabs or shellfish.

A killdeer, a year-round resident in the southern United States, wades through shallow water.

Killdeer

Charadrius vociferus
L 10.5"; W 3.3 oz   **F** **S**

appearance North America's best known plover, the killdeer has a brown back and wings, a white belly, and a white breast crossed by two black bands. In flight, its bright orange-buff rump is conspicuous.

food The killdeer feeds mainly on insects and earthworms. It is often seen foraging in man-made habitats, such as playing fields and parking lots. It also seeks food in a variety of wet areas, but rarely along the ocean shore.

behavior This species forages mainly by sight. It often runs in spurts at the edge of a freshwater pond or muddy lagoon. It is very vocal. The killdeer nests on barren ground.

range Breeds from southern Alaska across northern Canada to Labrador south across northern U.S.; migrates as far as South America; year-round resident across the southern U.S.

listen for the distinctive sounds for which the killdeer and willet are named: a far-carrying, excited *kill-deer* and a ringing *pill-will-willet*.

A willet
in flight

Willet

Tringa semipalmata
L 15"; W 8 oz **F** **S**

appearance The willet is a large, plump shorebird with grayish legs. It is plain gray above and pale below in winter; the breeding adult is heavily barred on its neck and breast. The striking black-and-white wing pattern is visible only during flight.

food Breeders on the northern Plains eat insects, while those breeding in eastern salt marshes prefer crustaceans, mollusks, and small fish. Winter birds forage on mudflats and coastal beaches.

behavior This bird nests in alkaline and freshwater marshes in the West and salt marshes in the East.

range Birds that breed in the East migrate to South America. Western breeders spend their winters on beaches and mudflats.

Greater Yellowlegs

Tringa melanoleuca
L 14"; W 6 oz **F** **S**

appearance This shorebird is a medium-size sandpiper with long, yellow legs; its bill is longer than its head. The similar lesser yellowlegs (not illustrated) is noticeably smaller and has a shorter, more needlelike bill.

food This species pursues small fish and insects at the water's edge or while wading in deeper water.

behavior The greater yellowlegs often forages quite actively, racing about with its neck extended. The call is a series of three or four loud, piercing *tew* notes.

range Nests in bog and open areas in Alaska and across Canada. Winters in coastal U.S. and some inland areas in southern states and is seen migrating throughout the U.S.

Greater yellowlegs

Spotted sandpiper

Spotted Sandpiper

Actitis macularius
L 7.5"; W 1.4 oz **F** **S**

appearance North America's most widespread breeding sandpiper has bold black spots on its underparts in breeding plumage; in winter, it has plain brown upperparts and is white below, with no spots.

food The diet includes insects, crustaceans, mollusks, small fish, and earthworms found near the water's edge.

behavior This bird constantly bobs and teeters while walking or standing. Its flight is low over the water, and its shallow wingbeats seem jerky. Females are polyandrous—they defend a territory and mate with several males, who take the lead parental role.

range Breeds across much of North America. Most birds migrate to Central or South America for the winter, but some remain in the southern U.S. and Pacific states.

Ruddy Turnstone

Arenaria interpres
L 9.5"; W 3.9 oz **S**

appearance This short-legged, chunky shorebird has a black-and-white pattern and some rusty patches, orange legs, and a black bill. It has a very striking pattern in flight.

food These birds feed on small mollusks and crustaceans, insects, and horseshoe crab eggs found while picking through rocks, shells, seaweed, and other flotsam.

behavior It uses its short bill to flip aside pebbles and shells in order to find prey, hence its name. Ruddy turnstones often occur in small flocks in the winter and during migration. The call is a low-pitched, guttural rattle.

range Breeds on Arctic coastal tundra. Winters along both coasts, south to South America. Rare inland, except in Great Lakes region during migration.

Ruddy turnstone

Sanderling

Calidris alba

L 8"; W 2.1 oz **S**

appearance The winter plumage is very plain—pale gray above and white below. Breeding plumage displays a variably rusty head, breast, and mantle. The juvenile, seen primarily in August and September, is similar to the winter adult but has a blackish, checkered back.

food The diet consists of marine invertebrates found in wet sand, as well as tundra insects in summer.

behavior A familiar sandy beach bird, the sanderling seemingly chases waves out to sea, only to be chased back by the next wave, its legs blurring in speedy retreat. It often occurs in flocks, except when breeding.

range Found along the Atlantic and Pacific coasts during most of the year; breeds on dry Arctic tundra from late May to early July. Some migrants travel as far south as southern South America.

Sanderlings

Bird Tracks and Signs

You can learn more about the creatures that frequent the water's edge by examining tracks and other signs, such as bird feathers. Taking photos of tracks and checking them with books or online guides will give you a more complete picture of an area's birds, mammals, and reptiles and how they fit in to the environment. On bird tracks, look at the length of the toes, whether or not they are webbed, and how far apart the steps are. If you identify a bird walking in a muddy area, you can later photograph its tracks and start a collection of known images. For scale, it helps to place a ruler in the photo frame.

Sanderlings, which are small wading birds, move with the ebb and flow of waves while feeding.

Gulls

Gulls belong to a family of web-footed waterbirds that includes terns and skimmers. Most gulls are large or medium-size gray-and-white birds with black markings when they are adults. The larger species take three or four years to reach adult plumage. Although commonly called sea gulls, many gulls live far inland.

Laughing gull

Laughing Gull

Leucophaeus atricilla
L 16.5"; W 11 oz **S**

appearance The summer adult has a smart black hood that becomes whitish in winter; long, black-tipped wings; and slate gray upperparts. The adult's bill is red in summer and black in winter. The juvenile has a brownish head and upperparts. It takes three years for a laughing gull to reach adult plumage.

food This gull feeds primarily on fish, insects, and crustaceans, but also scavenges for beach handouts and garbage.

behavior The laughing gull is very bold at public beaches and parking lots. Its name comes from its call, which sounds like a high-pitched laugh.

range Common on U.S. Gulf Coast and Atlantic coast, rare inland.

Ring-billed Gull

Larus delawarensis
L 17.5"; W 1.1 lb **F S**

appearance The adult's bill is yellow with a dark ring, and its legs are yellow; the plumage is white below and light gray on the back and wing coverts. First-winter birds have brownish wings, a gray back, and a dark tail tip. This species takes three years to reach adult plumage.

food These gulls feed on fish, insects, and small mammals, as well as garbage and waste grain. This notably adaptable feeder even soars to catch flying insects and plucks berries from trees!

behavior This species is bold around people and is often seen in flocks loafing on coastal beaches, riverbanks, fields, and parking lots.

range Common in large areas of North America, including inland. Nests mostly on islands in lakes—70 percent of the population nests in Canada.

Laughing gulls at sunset

The adult ring-billed gull has yellow eyes.

Herring Gull

Larus argentatus
L 25"; W 2.5 lb **F S**

appearance This large gull, larger than the ring-billed gull, has pink legs at all ages. The adult has a white head (streaked brown in winter), white underparts, a light gray back, and a yellow bill with a red spot at the tip. Immature birds are browner and quite variable. This species takes four years to reach adult plumage.

food The herring gull feeds on fish, invertebrates, other birds and their eggs, carrion, and garbage—just about anything.

behavior Some individuals have learned to drop large shellfish onto pavement in order to crack them open.

range Widespread in North America along coasts and large inland bodies of water. Common year-round along the East Coast and mainly a winter visitor along the West Coast.

Western Gull

Larus occidentalis
L 25"; W 2.2 lb **S**

appearance The adult has a white head and underparts, a dark gray back (much darker than that of the herring gull), and a heavy banana-yellow bill with a red spot near the tip. First-winter

A western gull landing

birds are mostly sooty brown with blackish bills. This species takes four years to reach adult plumage.

food The diet includes saltwater fish and invertebrates, carrion, and garbage. Like the herring gull, flying birds sometimes drop clams and crabs onto rocks to crack them open.

behavior Western gulls are most often seen loafing on beaches and piers. They breed in colonies on rocky islands, often near colonies of other seabirds or California sea lions, in order to prey on eggs or to scavenge dead animals.

range Abundant along the Pacific coast from the Canadian border to Baja California, Mexico, but rarely occurs inland.

Herring gulls

A Caspian tern displays territorial behavior at its nesting site along a riverbed.

Terns & Kingfishers

While you might see both terns and kingfishers diving for fish at the water's edge, they arrived there via different evolutionary paths. Terns are waterbirds related to gulls, although their bodies are lighter and more streamlined. Kingfishers are descended from land birds, and in other parts of the world, where there are many more species, most kingfishers are forest birds, such as the famous laughing kookaburra of Australia.

Caspian Tern

Hydroprogne caspia
L 21"; W 1.4 lb F S

appearance This is the world's largest tern (larger than many gulls). Unlike other terns, the adult has a stout, bloodred bill with black near the tip and extensive black wingtips. In the summer it has a jet-black crown that takes on a streaky salt-and-pepper appearance soon after breeding.

food This species feeds mainly on fish caught by plunge diving, but it also eats large insects, bird eggs, and young birds. Often flies long distances from breeding colony to find food.

behavior The call is distinctive, drawn-out, and raspy. Juveniles follow their parents for months and beg loudly for food with a high-pitched whistle.

range Nests in coastal areas, the Great Lakes, and other large inland bodies of water. Inland-nesting birds spend the winter months in coastal locations; some migrate as far as northern South America.

Forster's terns and belted kingfishers plunge-diving into the water from a height of ten feet or more to catch fish that they spied from a perch or while hovering.

Belted kingfishers

Forster's Tern

Sterna forsteri
L 14.5"; W 6 oz **F** **S**

appearance This bird is very similar to several other medium-size terns. Forster's tern has paler gray upperparts, an orange-red bill with a black tip, and pure white underparts. In winter, its black cap is replaced by what looks like a black bandit's mask, and its bill becomes blackish.

food Forster's terns feed mainly on fish caught by plunge diving. They also eat insects in summer.

behavior This species breeds in colonies in marshes. Its nest is a platform of reeds and grasses.

range Breeds along mid-Atlantic coast, across the northern U.S. from Great Lakes to Oregon, and along the Gulf Coast. Winters mainly along both coasts and nearby marshes.

Belted Kingfisher

Megaceryle alcyon
L 13", W 5 oz **F** **S**

appearance The belted kingfisher has a large head with a shaggy crest and a large, daggerlike bill. Its plumage is slate blue above and white below, with a single blue breast band on the male and an additional rusty band on the female.

food This bird feeds primarily on fish caught by plunge diving. Kingfishers need clear, relatively still water to see their underwater prey.

behavior This species is often conspicuous. It is seen perched on trees or posts and watching the water for fish. Pairs defend a linear territory along a stream, river, or lakeshore against other kingfishers. A loud, dry rattle call is often the first sign of its presence.

range Breeds near inland waters or coastal areas across North America; migrates to ice-free areas in winter.

Forster's terns breed primarily in coastal marshes.

did you know? **belted kingfisher parents** dig a six-foot-long, six-inch-diameter tunnel into a steep riverbank or sand bank to make a nest? The hole tilts upward to allow water to drain away.

Songbirds

Songbirds are small birds with vocal organs that have evolved to produce elaborate and diverse songs. In most species only the male sings a complex territorial song, but both sexes give short call notes. Many songbird species visit water's-edge habitats, but only a few are covered here. For more information, consult a birding field guide.

A common yellowthroat perches on a twig.

Common Yellowthroat

Geothlypis trichas
L 5"; W 0.4 oz **F**

appearance The adult male has a broad, black mask bordered above by gray or white, and a yellow throat and breast. The female is similar but lacks the black mask.

food These birds eat mostly insects, as well as spiders and occasionally seeds.

behavior Common yellowthroats usually stay well hidden in low vegetation but are very vocal when breeding. Listen for the male's *wich-i-ty, wich-i-ty, wich-i-ty* song coming from a thicket or marshy area. It nests on or near the ground in a thick clump of sedges or grasses.

range Nests across North America in summer and is a year-round resident in southern areas.

Tree Swallow

Tachycineta bicolor
L 5.8"; W 0.7 oz **F S**

appearance The adult is iridescent greenish blue above and pure white below, with a slightly forked tail. The juvenile is brownish above and has a faint breast band.

food Tree swallows eat flying insects, as well as some berries and fruit, mostly in winter.

behavior This species nests in natural cavities or nest boxes near open water and fields. It is found in large flocks outside the breeding season, especially during fall migration. It glides more than other swallows do.

range Breeds as far north as Alaska and northern Canada. Winters in the southern tier of states and along the eastern seaboard, where it subsists mainly on a diet of berries.

Tree swallows

Red-winged blackbird

Red-winged Blackbird

Agelaius phoeniceus
L 8.8"; W 1.8 oz **F**

appearance In the summer the male has glossy black plumage with red and yellow shoulder patches. In the winter it has glossy black plumage partially obscured by buffy feather fringes. Females and juveniles are very different—dark brown above and heavily streaked below.

food This bird eats insects, seeds, and berries; in winter it eats mostly grain.

behavior Red-winged blackbirds breed in wetlands in the summer, usually in colonies. Starting in early spring, the male displays his red patches and gives a hoarse, gurgling call—*conk-la-ree*—from a prominent perch.

range Breeds across North America, but retreats from coldest areas in winter and forms large flocks in rural areas.

Yellow-headed Blackbird

Xanthocephalus xanthocephalus
L 9.5"; W 2.3 oz **F**

appearance The adult male is black with a yellow head and breast and white wing patches. The adult female is brownish with a dull yellow throat and breast and lacks white wing patches.

Why Do Birds Migrate?

Some birds reside in the same area all year, but many species must migrate to survive, and some travel thousands of miles each year. The main reason is the reduced food supply in winter. Longer summer days favor the abundant growth of plants, insects, and other foods that many species rely on to eat and to feed to their hatchlings. Also, bitterly cold winters are difficult for many species to cope with. Decreasing daylight in late summer and early fall triggers the birds' migration instincts and sends them south.

food This blackbird eats aquatic insects, grain, and weed seeds. Breeding birds often forage in grasslands or farmland adjacent to their nest sites.

behavior The bird is conspicuous in many western wetlands, where it breeds in loose colonies. The nest is built over water and is attached to cattails or reeds. Each male defends a small territory within the colony by singing. The song is unusual—some would say unpleasant. It begins with a harsh, rasping note and ends with a long, descending buzz.

range Breeds mainly west of Great Lakes; most birds migrate to Mexico for the winter.

Yellow-headed blackbird

NEW JERSEY
Atlantic City
Cape May
N.W.R.
Dover
Delaware Bay
Cape May
DELAWARE
Atlantic Ocean
MARYLAND
0   20 mi.

RECOMMENDED
DR. BEACH
RECOMMENDED

ON LOCATION

Cape May, New Jersey

- More than the boardwalk
- Prime bird-watching
- Migration hotspot
- Swarms of horseshoe crabs

A flock of black skimmers takes flight near Cape May, New Jersey, a prime area for viewing birds.

Cape May, on the narrow peninsula between the Atlantic Ocean and Delaware Bay at the southern end of New Jersey, is widely known for its ocean beach, its boardwalk, and its collection of 600 late 19th-century buildings in the Victorian district. To bird-watchers around the world, Cape May and the Delaware shore across the bay are prime places for viewing and photographing migrating birds in the spring and

fall. Bird-watchers flock there in both May and October not only to see how many birds they can spot but also to take part in workshops and to take field trips on land and by boat to nearby marshes and other areas with expert guides organized by the Cape May Bird Observatory.

Cape May had been holding weekend and then weeklong birding events for 35 years when organizers switched to the monthlong CAPE

MAYgration Celebration in May 2012. Another May birding event is the New Jersey Audubon Society's World Series of Birding, a one-day competition in which participants identify as many birds as possible in New Jersey in 24 hours. Fall migrations also bring prime bird-watching to Cape May. In October 2011, the bird observatory held the 65th annual Cape May Autumn Birding Festival. It also offers workshops and other events for most of the year, such as weekly walks led by experts who focus on birds but also point out and discuss other aspects of nature.

Cape May Lighthouse in Cape May State Park

Birds and Horseshoe Crabs

Each year the biggest natural events on both the Cape May and the Delaware sides of the bay are the May and June invasions of beaches by horseshoe crabs intent on laying eggs and migrating birds intent on eating those eggs. For the rest of the year these arthropods, which are more closely related to spiders than they are to any crab, stay on the bottom of the oceans and bays. At the times of the May and June new and full moons, when tidal ranges are greatest, females with males hanging on their backs crawl up on beaches to lay eggs. The females scratch holes in the sand, lay 2,000 to 4,000 eggs that the males fertilize, and crawl back to the water. Females repeat this several times during the cycle. In their frenzy to lay eggs, females uncover some nests and expose eggs that are easy pickings for flocks of migratory birds. These include red knots that have flown nonstop from South America and need the eggs to fuel their flight to their Arctic nesting grounds. A decline the in numbers of horseshoe crabs seems to be reducing the numbers of red knots and could be affecting other long-range migrants. If you would like a close-up view of horseshoe crabs laying eggs, search on the Web for "Horseshoe crab census," which will take you to a website where you can volunteer to count arriving crabs for researchers.

See Migrating Raptors

While shorebirds often migrate long distances over open water, raptors such as hawks seem spooked by wide stretches of water. These birds act as though they are in a funnel bounded by the Atlantic Ocean and Delaware Bay as they head south, and this concentrates them around Cape May. They linger there until favorable winds allow them to take off for the 12-mile flight across the bay's mouth. The best place to watch them is the Cape May Hawk Watch platform on the ocean at Cape May Point State Park on New Jersey's southern tip. In their breeding areas, you rarely see more than one or two eagles, hawks, ospreys, or other raptors at a time. Here you can see hundreds. During migrations you'll usually find others at the platform to help you identify birds.

Whales are humanity's canary in the coal mine . . . As ocean pollution levels increase, marine mammals like whales will be among the first to go.

—CONSERVATIONIST ROGER PAYNE

[7]

MAMMALS

Dolphins, known for their agility and playfulness,
leap above the water's surface.

Manatees, Seals & Sea Lions

Like their land-based cousins, marine mammals are warm-blooded, use lungs to breathe air, and bear live young that they nurse with milk from mammary glands. They evolved from distant-ancestor land mammals with adaptations that make them efficient at swimming and able to stay underwater long enough to hunt and to keep warm in cold water.

West Indian manatee

months, and newborns weigh 60 to 70 pounds. The female nurses her calves for up to two years.

range Limited to tropics and subtropics.

West Indian Manatee

Trichechus manatus
L up to 15' **F** **S**

appearance This manatee is gray or brown. The forelimbs are flippers, and it has no hind limbs.

food The diet consists mostly of plants. Sea grasses growing on the seafloor are a major food source, along with some fish and small invertebrates. Adults eat 10 to 15 percent of their body weight daily.

behavior This species cannot tolerate cold water because of its low metabolic rate and lack of a thick layer of insulating body fat. It may live longer than 60 years in the wild. Its gestation period is 12 to 14

California Sea Lion

Zalophus californianus
L up to 8' **F** **S**

appearance The adult male is mostly dark brown, with a paler belly and side colors; the adult female is paler and sometimes tan. Pups are born dark but lighten after several months.

food This animal mainly eats squid and fish.

behavior Highly social, the California sea lion cooperates with dolphins, porpoises, and seabirds to hunt schools of fish. It adapts to man-made environments and is commonly

California sea lion

found in public displays in zoos and marine parks. It can remain underwater for up to 15 minutes and dive to 900 feet. Sea lions use echolocation to find food, to orient themselves, and to navigate underwater. Their main predators are killer whales and sharks.

range Mainly in California waters; overall from Alaska to Mexico.

Harbor seals resting on ice

look for **sea lions** that have learned to catch salmon and other fish ascending fish ladders at Bonneville Dam and other Columbia River dams.

Harbor Seal

Phoca vitulina
L up to 6' **F** **S**

appearance This seal is brownish black, tan, or gray with paler underparts, short flippers, a large rounded head, and a V-shaped snout.

food The diet includes fish, crustaceans, mollusks, and sometimes seabirds. The seal often swallows its food without chewing.

behavior A layer of fatty blubber under the skin helps the harbor seal to maintain its body temperature. It can dive to 1,450 feet and stay submerged for 25 minutes. These seals often return to the same resting site after swimming more than 30 miles in search of food. Many swim far upstream in large rivers. It moves on land by bouncing along on its belly; it rests on rocky coasts, sandy beaches, muddy areas, and ice. Females usually bear a single pup, which they care for alone. Pups are able to swim and dive within hours of birth.

range Temperate and Arctic coasts from Greenland and Canada to New Jersey on the Atlantic coast, Alaska to southern California on the Pacific coast.

Protecting These Animals

The U.S. Marine Mammal Protection Act, passed in 1972, prohibits "taking" any marine mammal in the United States, taking a marine animal anywhere in the world if you are a U.S. citizen, and importing marine mammal products. "Taking" includes harassing, collecting, capturing, or attempting to do any of these. The law has exceptions allowing Alaskan native people to continue their traditional hunting of animals such as seals and some whales. Scientists must obtain permission for research projects.

Both U.S. and Florida law protect manatees.

Many people know bottlenose dolphins as entertainers.

Dolphins

Greek, Hindu, and Native American mythologies include dolphins. They are stars at today's marine parks, in the 1960s *Flipper* movie and television series, and in the 1990s *Free Willy* movies. Their natural lives are, if anything, more fascinating than those described in ancient and contemporary myths.

Bottlenose Dolphin

Tursiops truncatus
L up to 13' **S**

appearance The upper body is light gray to slate gray, shading to paler on the sides; the belly is pale, pinkish gray. The dorsal fin is high, curved, and near the middle of the back.

food The diet includes eels, squid, shrimp, and fish; the animal swallows its food whole.

behavior These dolphins typically live in pods of about 15, but pods can be as large as 1,000, and some individuals are solitary. Groups often work as teams to hunt fish. To search for prey, they use reflected sound waves from clicks they emit. They communicate with squeaks, whistles from their blowholes, and sounds made by body movements such as slapping tails on water. The gestation period is 12 months, and calving occurs year-round; calves nurse for 12 to 18 months.

range Temperate and tropical oceans worldwide.

look for **bottlenose dolphins** jumping and playing in the surf. Sometimes you will even see them riding waves much like a human bodysurfer.

Common Dolphin

Delphinus delphis
L up to 8′ **S**

appearance This dolphin's back is dark gray to black from the top of head to the tail. The flanks are tan to yellowish tan in front of the dorsal fin but light gray behind the fin. The belly is white or off-white. A dark band connects dark circles around the eyes.

food It eats any small species of fish, as well as squid.

behavior The common dolphin is very social. It often travels in pods of more than 1,000 and can be seen riding the bow waves of ships. The male is sexually mature in 3 to 12 years, the female in two to seven years. This species breeds in summer and has a 10- to 11-month gestation period. The female gives birth to a single calf and nurses for four months.

Common dolphin

range Mostly on continental shelf edges, Atlantic from Newfoundland and Nova Scotia to northern South America, Pacific from Victoria, British Columbia, to Equator.

Killer Whale

Orcinus orca
L up to 26′ **S**

appearance The killer whale is black above and white below, with a white patch above and behind the eyes. It has a variable gray or white saddle behind the dorsal fin. The male is larger than the female.

A killer whale leaps from the water.

food Different populations specialize in particular kinds of prey, including fish and other marine mammals.

behavior Like other dolphin family members, the killer whale is highly social. Some populations consist of family groups that pass on hunting techniques. It uses sophisticated sounds for navigation, hunting, and communication to new generations. The female has one offspring, roughly once every five years. Both the male and the female care for their young.

range World's oceans; most common in the Arctic and Antarctic, and often seen off the North American West Coast.

Whales

All whales are one of two types. Toothed whales, such as dolphins, grab prey with their teeth and swallow it whole or bite off pieces to swallow. Baleen whales have plates hanging from their upper jaws. They take in huge gulps of water. When they expel the water, the baleen plates filter out food, ranging from plankton to small fish, which the whales swallow.

A humpback whale shows its flukes as it dives.

Humpback Whale

Megaptera novaeangliae
L up to 50'   S

appearance The humpback has a mostly stocky body, with white on the throat grooves. The flippers, which are one-third as long as the body, are white below and mottled black and white above. The tail flukes are long.

food The diet is mostly shrimplike krill and small fish. It uses baleen to feed.

behavior This species spends its summers feeding in far northern or far southern waters, and then it migrates as far as 16,000 miles to warmer water to breed and to give birth. It lives off fat reserves until the spring migration to colder water. Individuals live alone or in small groups, usually for a few hours, to cooperate in foraging.

range Worldwide.

Whale-watchers see a humpback's long fins.

Gray whales coming up for air

Gray Whale

Eschrichtius robustus
L up to 45' **S**

appearance This whale has a streamlined body and a narrow, tapered head. The body is slate gray with gray patches and white mottling on dark skin. There is no dorsal fin. It has 6 to 12 bumps on the middle of the rear quarter. Flukes are 10 to 12 feet across, and flippers are broad, paddle shaped, and pointed at the tips.

food It scoops up and eats ocean-bottom crustaceans.

behavior North American whales feed in the Bering and Chukchi Seas off Alaska in the summer. In October they begin migrating 5,000 to 6,800 miles in two to three months along the West Coast to warm water off Baja California or the southern Gulf of California. Gestation takes a year; calves are born in shallow lagoons. Northward migration takes two to three months; mothers and calves stay close to shore.

range North Atlantic, western Pacific.

look for **roughly one-third** of the world's whale and dolphin species near and around Channel Islands National Park in California.

Meet the Whales Whale-watching gives you a brief look at these huge, fascinating animals in their habitat. Many places along North America's Pacific coast offer good views of migrating whales from high vantage points that allow you to watch several whales surface and dive. For close-up views, you can choose from scores of whale-viewing boat trips in Alaska (in the summer) and other places along the Pacific coast, and from Virginia north along the New England and Canadian coasts. If you want to use your own boat for an intimate look at whales, begin educating yourself by going to the whalesense .org website, which hosts a voluntary education program for commercial whale-watching companies. Following this site's guidelines will keep you from violating U.S. marine mammal protection laws, hurting yourself, and disturbing whales. Before buying tickets for a whale-watching cruise, find out whether the operator follows the guidelines. Before you board a cruise, be prepared for hot sun, cooler temperatures, and splashes from waves or whales. If motion sickness is a concern, be sure to come prepared.

Whale-watchers get a close-up view.

Sacramento
San Francisco
San Jose
Monterey Bay N.M.S.
Carmel   CALIFORNIA
Hearst Castle Cambria
Piedras Blancas Beach
Los Angeles
Pacific Ocean
0   100 mi.
San Diego

ON LOCATION

Piedras Blancas Beach, California

- Thousands of elephant seals
- No admission charge
- Easy drive from cities
- Stunning coastal scenery

This California beach story reads almost like a television reality show. A few big guys show up at a beach one by one in November, sprawl on the sand for a while, and begin challenging one another with grunts and snorts. Eventually, two of the guys bang their chests into each other and make more grunts. Others engage in their own fights, which include bloody

Looking north along Piedras Blancas Beach, ice plants in the foreground

biting but rarely end in death. Finally, one backs away while making weaker grunts. The battles continue through December, when a few alpha males become rulers of patches of beach. Then the females arrive and form harems around the alpha males. These big guys are really big: two-ton, 14- to 16-foot-long male elephant seals. Each harem has 30 to 100 females, which are 10 to 12 feet long and weigh 1,200 to 2,000 pounds.

Pull in and Watch

Piedras Blancas Beach is roughly 200 miles south of San Francisco or 270 miles north of Los Angles on California's Pacific Coast Highway. It is about seven miles north of San Simeon. To watch the seals, pull in to the beach's parking lot between the highway and the dunes. It is free and open every day of the year. Signs explain the seals' lives, and you can look for a Friends of the Elephant Seal docent in a blue jacket to answer questions. The National Marine Fisheries Service, which enforces the Marine Mammal Protection Act, says you should stay at least 50 to 100 feet from any seal, and if you notice that a seal is looking at you, you should back off.

Winter is the best time to watch the seals because the males are fighting

Elephant seals on Piedras Blancas Beach

and the females are having their pups, which were conceived the previous year. After giving birth, the females nurse their pups for about a month as the little ones grow from approximately 75 to 250 pounds. After her pup is weaned, a female mates and returns to the ocean. The pups hang around for a couple of months and teach themselves how to swim before leaving to begin their lives of diving to depths of 1,000 to 2,000 feet for food. The last adults usually leave Piedras Blancas by March.

More Than Seals

From late March until late November, when adult males begin to return to fight and mate, males and females of various ages return to molt—to shed the outer layer of skin and to grow new skin. While the seals are doing little except molting or resting from the labor of hunting under the ocean, you will find plenty of wildlife to view along the Central Coast of California, which is part of the Monterey Bay National Marine Sanctuary. It runs from Cambria, a few miles south of Piedras Blancas Beach, to just north of the Golden Gate Bridge. The sanctuary includes several areas for watching birds or marine mammals, including the California Sea Otter Game Refuge near Carmel, north of Piedras Blancas. When you want a break from the natural world, you can visit the lair of another alpha male not far from Piedras Blancas. This is Hearst Castle, which is atop a hill in the Santa Lucia Range, 1,600 feet above the ocean. William Randolph Hearst, the newspaper magnate, had the castle built between 1919 and 1947. It is now a state historic park that offers a variety of tours.

Nutrias are damaging saltwater and freshwater wetlands.

Rodents

All rodents, such as nutrias and muskrats, have teeth and muscles used for chewing and are adapted to eating plants, although some also eat animals. Almost all rodents are small; the largest is the American beaver, which can weigh more than 50 pounds. Raccoons are not rodents. They are carnivores, the large group of meat-eating animals that includes cats and dogs.

Nutria

Myocastor coypus
L up to 25" **F** **S**

appearance A nutria's upperparts are dark yellowish brown or reddish brown, masking slate gray underfur. It has large, yellow to dark orange front teeth. The forelegs are small compared with its body size, while the hind legs are large compared with the forelegs.

food This rodent predominantly feeds on the bases of plant stems. It digs for roots and rhizomes in winter and sometimes eats small animals.

behavior The nutria breeds year-round,

with 1 to 13 young in a litter. This animal is more at home in water than on land. It feeds mainly before sunrise and after sunset. It burrows in levees, dikes, and embankments—which means it often damages wetlands.

range Native to South America, domesticated as a fur animal and introduced to North America in 1930s. Fur farm escapees are considered invasive in a dozen U.S. states and parts of Canada.

look for **erosion in salt marshes,** caused by nutrias burrowing and eating plant roots, thus allowing tides to wash away fragile organic soils.

safety tip

Avoid any raccoon that does not fear you or is acting strangely. It could have rabies, which is common among raccoons.

Northern Raccoon

Procyon lotor

L up to 36" **F** **S**

appearance This raccoon is gray to reddish brown to buff and has a black mask across the eyes and a bushy tail with four to ten black rings. Its forepaws resemble human hands.

food The diet includes crayfish, insects, other invertebrates, rodents, frogs, fish, bird eggs, and sometimes plant life.

behavior Raccoons easily adapt to many habitats, including those near humans, though it needs ready access to water. It builds dens in trees, woodchuck burrows, caves, mines, and buildings. One litter of three to seven young is born each year. Young are born blind and helpless; their eyes open at 18 to 24 days.

range Across southern Canada, most of U.S. into northern South America

Northern raccoon

Common muskrats live in wetlands.

Common Muskrat

Ondatra zibethicus

L up to 23" **F** **S**

appearance The dense, glossy fur is dark brown above; lighter on sides; and finer, softer, and paler below. The throat is almost white, and the long tail is scaly and nearly naked.

food Muskrats eat mostly aquatic vegetation such as cattails, sedges, rushes, water lilies, and pond weeds; in some areas they eat freshwater clams, crayfish, frogs, and fish.

behavior This species is most active at dusk, dawn, and night. An excellent swimmer, propelled by slightly webbed hind feet, it spends most of its time in water. Using its tail to steer, it travels great distances underwater. It usually tows food to a feeding platform littered with plant cuttings and other scattered food debris. The muskrat builds small lodges similar to beaver lodges. It often damages dams by tunneling.

range From northern North America to Mexican border.

Beavers & Bats

Beavers and bats are remarkable examples of the evolution of animals over millions of years to live in unique environmental niches. Beavers are freshwater aquatic mammals, not marine mammals. Unlike marine mammals that hunt in the oceans, beavers find most of their food on land, but the water environment they create with their dams protects them from predators while giving them easy access to the land plants they eat. The evolution of wings gives bats access to flying insects and daytime sanctuary—in caves or perhaps in your attic—from birds such as hawks. The use of reflected sound to find food enables bats to hunt at night when hawks, which rely on keen eyesight, are not flying.

American beaver

A beaver building a dam

American Beaver

Castor canadensis
L up to 3' **F**

appearance The American beaver is dark brown, with a massive head, small ears and eyes, four large orange incisors, and a flat, scaly tail.

food Beavers eat bark; the cambium layer under bark; twigs, leaves, and roots of deciduous trees such as alder, willow, birch, and aspen; many parts of aquatic plants; and young water-lily shoots.

behavior This species mates for life. Three or four kits, born in May or June, begin swimming in a week. The flat, paddle-shaped tail and large, webbed hind feet make it a good swimmer. It slaps the water's surface with its tail to warn of danger. Beavers build dams to form ponds, and they also construct dome-shaped lodges with underwater entrances and air holes on top.

range Most of North America.

A beaver dam and the pond it created

gather in day roosts. They give birth and care for young in tree holes, other natural cavities, and attics. Offspring are born blind, and the eyes open in two days; the mother can fly with young attached to a nipple. Young can fly three weeks after birth, and they reach adult size in roughly a month.

range Alaska, across Canada and most of U.S. to central Mexico, mainly in forested areas.

Little Brown Bat

Myotis lucifugus
L up to 4" Wingspan up to 11" **F** **S**

appearance This bat is uniformly dark brown with glossy fur on the back; the underside is paler, usually with a grayish tinge. Wing membranes are typically dark brown.

food The diet includes moths, wasps, beetles, gnats, mosquitoes, mayflies, and other insects with an aquatic life stage.

behavior Fast flight, plus rapid chewing and digestion, enable a bat to eat as many as 600 insects in an hour. This species prefers to roost near water In buildings, trees, under rocks, in wood piles, or in caves during the day; after the evening feeding, large numbers

look for **bats zipping** back and forth, up and down, just above a lake, pond, or stream, as they hunt insects just before and after sunrise or sunset.

Little brown bat

did you know? **beavers increase** biodiversity by building dams that create ponds and wetlands where invertebrates, fish, mammals, and birds colonize? The ponds also remove pollutants and improve stream water quality.

Otters

Otters are members of the family Mustelidae, which also includes weasels, skunks, and badgers. Members of the family are active all year. Otters' fur is thick, which is why sea otters had been nearly pushed to extinction by 1911, when the United States, Great Britain, Japan, and Russia signed a treaty restricting seal and sea otter hunting. Other members of the family, including mink, ermine, and sable, are farmed and hunted for their fur.

Sea otters live on the Pacific coast.

Sea Otter

Enhydra lutris
L up to 4' **S**

appearance The fur is usually deep brown, with silver-gray speckles; an adult's head, throat, and chest are paler than the rest of the body.

food Sea otters eat marine invertebrates, including sea urchins, mollusks, and crustaceans, as well as some fish.

behavior This animal can walk on land but lives mostly in ocean. Unlike other sea mammals, it has no insulating fat; dense fur traps air to insulate the otter from chilly water. It uses long, highly sensitive whiskers and front paws to find prey in murky water. This is the only otter species to give birth in water.

The mother holds infants in order to nurse them.

range North American Pacific coast from Alaska to southern California; parts of Russian east coast.

Northern River Otter

Lontra canadensis
L body up to 30"; tail up to 17" **F**

appearance The fur is dark brown to almost black above and paler below. The throat and cheeks are usually golden brown. It has a long, streamlined body; a thick, tapered tail; short legs; a wide, rounded head; and small ears.

food The diet consists of mainly aquatic organisms, including amphibians, fish, turtles, crayfish, crabs, and other invertebrates; it sometimes eats eggs, birds, small terrestrial mammals, and aquatic plants.

A northern river otter feeds on fish.

A raccoon's unique tracks

Identifying Tracks

Your water's edge explorations will be more satisfying if you learn about the animals you might see, including what their tracks look like. Sand or mud near water shows the tracks of animals that visit the area. When you see an animal's tracks, you can try following them to see where the animal went. You can also look for other signs, such as plants that an herbivore has chewed or remains of a carnivore's dinner. Take photos of tracks you do not recognize so that you can look them up. To begin your education, try searching for "animal tracks" on the Web. The information you find will give you some insight into animals' lives, and it might help you to find more tracks on future explorations.

behavior The northern river otter lives in freshwater and coastal marine habitats, including rivers, lakes, marshes, swamps, and estuaries. It disappears from areas with polluted waters. This species builds its den in the burrows of other mammals, in natural hollows, and under logs. The den has an underwater entrance—a tunnel leading to a nest chamber.

range Throughout Canada and U.S. except in dry areas of southern California, New Mexico, Texas, Nevada, and Colorado.

American Mink
Neovison vison
L up to 22" F S

appearance This animal's long, streamlined body is light to dark brown, with a white to cream underside and a large, bushy tail. Winter fur is generally dark blackish tawny to light tawny; summer fur is generally shorter and duller.

food This mink eats muskrats, rabbits, mice, chipmunks, fish, crustaceans, snakes, frogs, and birds.

behavior American minks kill prey by biting it on the neck, swim and dive with ease, and can remain underwater for many minutes. The mating season is from January to April; the female has a litter of three to six young in April to May; babies are weaned at five to six weeks, but they stay with their mother until fall. Individual animals defend territories, and male-female territories overlap. This mink prefers rocky habitats near water with dense cover. Its dens consist of burrows in riverbanks, under logs, and in hollow trees.

range North America from Alaska and Canada across the U.S. except for arid parts of the Southwest and California.

American mink

ON LOCATION

Barrow, Alaska

- On the Arctic Ocean
- Easy to reach by airline
- A bird-watching paradise
- Under the midnight sun

Barrow, Alaska, is the only easy-to-visit polar ocean water's edge in North America. Instead of going on an expedition or a $10,000 (or higher) cruise across the top of North America via the legendary Northwest Passage, you take a flight (no roads go there), stay in a comfortable hotel, and eat at your choice of several restaurants. Barrow, the northernmost U.S. town, is 329 miles north of the Arctic Circle on the Arctic Ocean.

No Trees, Flocks of Birds

As you fly into Barrow on an Alaska Airlines Boeing 737, you see an expanse of green dotted with ponds but no trees. For much of the year

A mother polar bear and two cubs on the Arctic Ocean beach at Barrow

Arctic Ocean sea ice snuggles up to Barrow's beach, but it retreats during the summer except when north winds occasionally push it back. Summer bird-watchers can view some of the 250-plus bird species that nest on the tundra in the area, which is considered one of the best bird-watching places in Alaska, if not in the country. If you just cannot break away from the birds, you can watch them 24 hours a day between May 10 or 11 and August 2 or 3, while the sun never sets.

An Inupiat man on the sea ice near Barrow

A Different Natural World

In the summer, the verdant land is tundra, a thin layer of soil on top of permafrost, which has remained continuously below 32°F for more than half a million years around Barrow. Since water cannot soak far into the ground, the tundra is mushy when you walk on it, and it is covered with ponds. The soil atop the permafrost freezes each fall and begins thawing in early summer. Only the top 20 inches or so around Barrow thaws, which means plant roots cannot grow deep enough for trees.

Nevertheless, the area is far from barren. Biologists have identified 124 species of vascular plants—many with small flowers—plus sedges, grasses, rushes, a few low shrubs, 177 species of mosses, and 49 species of liverworts. Land mammals include migrant caribou, arctic and red fox, and lemmings. The Arctic Ocean's rich food webs support polar bears, walruses, whales, and several species of seals, as well as fish. Arctic tundra supports millions of nesting birds because the summer explosion of life under the midnight sun supplies plenty of food, such as insects, including big mosquitoes.

The region's most impressive bird is the snowy owl, with a wingspan up to 62 inches. The champion flier is the bar-tailed godwit. From August 29 to September 7, 2007, U.S. Geological Survey biologists used a surgically implanted satellite transmitter to follow a bar-tailed godwit's nonstop, 7,250-mile flight over the Pacific Ocean to New Zealand. (These birds do make some stops on the way back in spring.) October, after many birds have flown away to the south, is the best time to see marine mammals. Bowhead whales, which Inupiat Eskimos hunt as they swim past Barrow in the spring and fall, as well as beluga and gray whales, pass Barrow on their annual migrations. You are most likely to see polar bears from October to June.

Be Ready for the Weather

If you visit Barrow in the summer, a jacket, a hat, and lightweight gloves should be fine. July's average high temperature is 47°F, and the low is 34°F. If you go in October, be ready for cold; on October 1, the average high and low temperatures are 33°F and 22°F. By October 31, the average high and low are 13°F and 3°F. On January 30 the average high and low are minus 8°F and minus 20°F.

*We need the tonic of wildness to wade
sometimes in marshes . . . to smell the
whispering sedge where only some wilder
and more solitary fowl builds her nest.*

—HENRY DAVID THOREAU, *WALDEN*

[8]

PLANTS

A river and a marsh meet in a kaleidoscope of color.

Algae & Duckweed

With a three-billion-year-old pedigree, algae occur in seas and fresh water in forms ranging from a single cell to complex, multicellular organisms. Algae photosynthesize—which long caused scientists to group them with true plants—and provide food and oxygen to other organisms. Duckweed, a true plant, shares its small size and rapid spread with some algae.

Green algae at a Yellowstone hot spring

habitat Green algae are primarily found in aquatic habitats, although they also occur in a wide variety of land habitats. They live in all types of water—fresh, brackish, and salt.

range Widespread throughout North America and worldwide, forming the basis of the food chain for most aquatic life. An unfettered algal bloom can also extinguish other aquatic life.

Plankton

various species
L microscopic to 100′ [F] [S]

appearance This varies widely, as plankton include microscopic single cells, animals, plants, and bacteria that drift near the water's surface and form the basis of the marine food web. Some organisms are temporarily planktonic, such as during the egg or larval stage. Lion's mane jellyfish, at 100 feet long and 8 feet wide, is considered the largest plankton.

reproduction Methods include simple cellular division, asexual budding, and sexual reproduction.

habitat Plankton are found in ponds, lakes, seas, and oceans to varying depths depending on sunlight and nutrients.

range Widespread, occurring in freshwater and saltwater environments throughout North America.

Green Algae

various species
L microscopic to macroscopic [F] [S]

appearance The 7,000 species of green algae vary widely in appearance, from single cells to colonies to long filaments and seaweeds. The green color comes from chlorophyll, a pigment vital to photosynthesis.

reproduction These species reproduce by numerous sexual and asexual methods, including fusion, fertilization, and cell division.

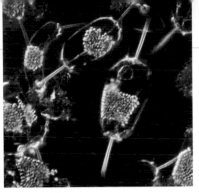

A group of plankton

Pithophora

Pithophora species
L microscopic   **F**

appearance This lime green to brown, irregularly branched, filamentous green algae grows in clumps forming dense mats on the bottom or surface of the water. Gas bubbles given off by plants make the mat buoyant. These species resemble horsehair, soggy wool, or cotton, which has given rise to several common names, including horsehair or cotton-ball algae.

reproduction These organisms reproduce by cell fusion of identical gametes, which the algae produce in large quantities.

habitat Pithophora are found exclusively in freshwater habitats such as shallow ponds and lakes, especially those with high nitrogen and phosphorus content.

Pithophora, a type of green algae

range Occurs throughout North America. Grows prolifically, covering acres of surface.

Duckweed

Lemna species
W 0.25"   **F**

appearance The world's smallest flowering plant, duckweed is composed of two tiny bright green leaves and a single root that hangs below the water's surface. It usually occurs in dense colonies. Not all species flower, but of those that do, the flower has two stamens and one style.

A garter snake in duckweed

reproduction This organism reproduces vegetatively, with new plants growing from leaf buds on "mother" plants. The flower is about 0.04 inch wide.

habitat Duckweed occurs in nutrient-rich, still or slow-moving freshwater habitats. It is also produced commercially as feed in fish farming.

range Found throughout North America, except in some far northern areas. Duckweed is often transported to new locations by waterfowl.

Seaweeds

Encountered on beaches as clumps of tide-borne detritus, seaweeds represent many large and complex red, green, and brown marine algae. They mostly inhabit shallower coastal waters and occupy many ecological niches, but ultimately they are dependent on adequate sunlight for photosynthesis.

A kelp perch swims among giant kelp off the Pacific coast.

Kelp

various species
L to 240' **S**

appearance A kind of brown algae, long streamers of kelp attach to the seafloor by means of rootlike structures called holdfasts. Leaflike blades grow on stems, kept buoyant by gas-filled bladders.

reproduction Specialized kelp blades shed spores, which germinate and create male and female gametophytes. Fertilization creates new organisms.

habitat Large colonies of kelp, called forests, form in clear, cool, nutrient-rich shallow waters off rocky coasts. In the Pacific, sea otters use kelp strands to anchor themselves while feeding.

range Kelp forests found mainly off North Atlantic and Pacific coasts. Range in depth from 20 to 120 feet, depending on amount of light available for photosynthesis.

Knotted wrack with fruiting bodies

A handful of seaweed

Gathering to Eat

For millennia, humans have combed coasts to gather seaweeds for use as food. The nutritious algae contain important vitamins and minerals, and they can be eaten raw, dried, and cooked. Seaweed foragers should abide by all collection laws, know how to identify species, and try to take only plants that already are detached. Except for tossings from new storms, avoid beached plants, which have started to decay.

Knotted Wrack

Ascophyllum nodosum
L 3-9' **S**

appearance Also known as rockweed or bladder wrack, this tough, brown seaweed is formed from narrow fronds with oval gas bladders and attached to rocks by holdfasts. Each year it forms one new bladder along the frond, allowing age approximation.

reproduction The female plant grows conspicuous, raisinlike egg receptacles. Eggs are released from them, and the male alga adds sperm from its receptacles.

habitat This species is found primarily in cool waters on rocky shores in sheltered intertidal zones.

range Occurs along coast of North Atlantic in Canada and U.S. Found sporadically off San Francisco, probably as a result of use as shellfish packing. Harvested commercially.

Irish Moss

Chondrus crispus
L 3-7' **S**

appearance This is a red alga with fanlike fronds, which may be curled. It ranges from white to green to dark red in color. It is a source of carrageenan, a gummy substance commonly used to thicken foods and to hold liquids in suspension.

reproduction Irish moss reproduces like other algae do, but it has a third stage: In addition to sporophytes and gametophytes, it has carpospores.

Irish moss is a kind of red algae.

habitat This species lives in tide pools and often as thick carpet on rocks and ledges in shallow tidal areas. It is also found on cobble shores. In estuaries it can occur to a depth of 72 feet.

range Found mainly on North Atlantic coast from Labrador to New Jersey and on California coasts. Commercially produced for use in manufacturing.

Shallow-Water Sea Grasses

Sea grasses do what other flowering plants cannot do: The majority live and reproduce entirely underwater. They form vast meadows in shallow coastal waters, but they are not true grasses; they have more in common with lilies and gingers on land. But sea grasses' supple stems let them undulate with currents and tides. Sea grasses feed marine animals and waterfowl and serve as nurseries for many species.

Eelgrass

Zostera marina
L 6" to 3' S

appearance Bright green, quarter-inch-wide leaves grow in dense clumps resembling tangled ribbons. Eelgrass beds are usually completely submerged. Plants growing in sand have narrower leaves than those of mud-rooted ones. When uprooted, eelgrass turns brown or black.

reproduction This plant reproduces by rhizomes and also by seeds, which float trapped in the dense vegetation until they fall free and settle on the substrate to germinate.

habitat Eelgrass grows on mud or firm sand, and it forms extensive meadows in sheltered, shallow coastal waters. Eelgrass and turtle grasses are favorites of manatees, which graze in Florida waters and can eat a tenth of their weight in plants—about 100 pounds—each day.

range Found in Pacific and North Atlantic coastal waters as far south as South Carolina. In warmer latitudes, grows more prolifically in cooler months of spring and fall and dies back in summer.

Low tide exposes eelgrass on the Pacific coast.

did you know?

sea grasses sometimes are confused with seaweeds, despite their differences? One difference is that sea grasses send true roots into the sediment, while seaweeds use rootlike hold-fasts to grab firm surfaces.

Wet sea grass droops over rocks in a tide pool.

look
for

turtle grass beds when sailing near the coast in clear waters. They anchor sediment and help prevent erosion during hurricanes.

Turtle Grass

Thalassia testudinum
H 4–30" **S**

appearance Flat, straplike blades with parallel veins rise above an extensive root system. This grass forms extensive beds, essential to coastal stability. The blades are often covered with a variety of tiny organisms that use their host for support, not food.

reproduction Turtle grass reproduces by rhizomes and also by seeds formed in flowers that vary from white to pink.

Scouler's Sea Grass

Phyllospadix scouleri
H to 6' **S**

appearance This sea grass's long, narrow, flattened, emerald green blades with three veins grow in a mass up to six feet from a prolific, tuberlike rhizome base. Its short spadix, or flower stalk, bears two rows of flowers enclosed in a leaf sheath.

reproduction This plant reproduces sexually, with male and female flowers on separate plants. Female flowers form one-seeded fruits. Seeds are dispersed by waves. It also reproduces by creeping rhizomes, with which it anchors to a substrate.

habitat Scouler's sea grass forms large, single-species meadows found in tide pools or low intertidal levels on rocky shores, frequently exposed and hardy to wave action. It is sensitive to environmental pollution, such as oil and sewage spills. This species provides a habitat for the spiny lobster.

range An abundant Pacific coast species related to eelgrass. Ranges from Alaska south to Baja California.

Submerged turtle grass

habitat This grass grows on a variety of substrates in sheltered waters from the low-tide zone to depths of 90 feet. It significantly enhances the substrate depth by trapping sediments in its expanse of dead leaves and rhizomes.

range Found subtropically along the Atlantic coast south from central Florida and along the Gulf Coast.

A sand path leads through North Carolina dunes stabilized by grasses.

Dune Plants

Dune plants stabilize sand dunes—dynamic barriers to wave and wind action that protect the low-lying terrain of coasts and shores. Grasses, forbs, and shrubs take hold tenaciously in the nutrient-poor sand. Animals, from insects to mammals, shelter in the hardy vegetation.

American Beachgrass

Ammophila breviligulata
H 2-3' **F** **S**

appearance Upright, dark green, leafy stems grow in thick clumps. Stems are topped with dense, cylindrical, spiky seed heads about ten inches long.

reproduction Beachgrass spreads laterally and rapidly by vigorous runners. New growth can penetrate four feet of deposited sand. It also reproduces sexually for seed formation.

habitat This plant lives on beaches and dunes. Dunes cannot withstand foot traffic or off-road vehicle traffic, which most coastal states discourage or prohibit. Stick to defined paths through the dunes.

range Found along North Atlantic coast to North Carolina and along shores of the Great Lakes. Forms a dominant species along northern Atlantic. Used extensively in dune-restoration projects; effective from first season.

Virginia glasswort in fall bloom

Chickadee on a bayberry branch

Dune Plants & Wildlife

Plants sustain wildlife on dunes and beaches. Their leaves, flowers, and fruits provide food for many species. They offer shelter from strong winds and stinging sands, a secure place for birds and other animals to nest, and cool sands for burrowers. Threatened species often depend on the presence of dune vegetation—and dunes themselves.

Virginia Glasswort

Salicornia depressa
H 10-16" **F** **S**

appearance Waxy, scalelike leaves wrap around branched, fleshy stems the width of a pencil. Flowers occur in groups of three embedded in stem joints. The plant stays green through summer and turns from yellow to orange to red in autumn.

reproduction This species of glasswort is an annual that reproduces by seeds fertilized by open pollination.

habitat This is a distinctive plant found in salt marshes and on tidal flats; it is easily identified in contrast to marsh grasses.

range Found on Atlantic, Gulf, and Pacific coasts and shores of Great Lakes. Name stems from its use providing soda ash when burned to promote melting in the glassmaking process.

Sea Oats

Uniola paniculata
H 6' **F** **S**

appearance At maturity, this sturdy-stemmed perennial green grass bears long seed spikes with gently drooping spikelets. These turn strawlike in autumn. Deep and massive root structure stabilizes sand.

reproduction Spreads laterally by rhizomes; also spreads by windborne seeds dropped from the spikelets.

Sea oats, a common North Atlantic beach plant

habitat Sea oats grow on coastal dunes and along beachfronts. They are a dominant, unmistakable beach plant species along the southern Atlantic coast. They can also be purchased at nurseries for deliberate planting.

range Found from mid-Atlantic coast south to Florida and along Gulf Coast. Frequently planted in erosion-control programs and protected by law.

Shrubs

Sturdy, dense, and resilient, coastal shrubs anchor sand and soil and temper the effects of relentless wind. Additionally, many offer a bounty of tempting flowers and fruit for animals—including humans—along the shore. Some of these shrub species grow to tree form in less challenging environments.

Beach roses in bloom in Rhode Island

habitat The plant grows on dunes and sandy and rocky beaches. It is often cultivated and spreads readily. Despite its popularity, the beach rose competes with native vegetation and ranks high on lists of unwelcome invasives.

range An introduced species from East Asia, the beach rose is naturalized in northeastern North America and in the Pacific Northwest.

Beach Rose

Rosa rugosa
H 1–6′ **F** **S**

appearance This low shrub has wrinkled, compound green leaves and spiny stems. Large, sweetly fragrant single or double blossoms range from white to yellow to pink to magenta. Its prominent rose hips look like cherry tomatoes. This species is also known as the Japanese rose.

reproduction The beach rose reproduces sexually and creates abundant seeds, which often are dispersed by the animals that consume them.

Beach Plum

Prunus maritima
H 4–18′ **F** **S**

appearance This shrub has hairy stems and rough-surfaced, toothed leaves. Clusters of white flowers turn pink when pollinated. The round fruit is red to dark purple, sometimes yellow.

reproduction The flowers are pollinated by bees or wind. The fruit bears a single, swollen, egg-shaped seed. The plant also sends out suckers.

Beach plums

habitat The beach plum is native to sandy shores and dunes. It also thrives inland in varied soil types, with the exception of clay. Inland trees can grow to over 18 feet. It is propagated commercially and used to make jam.

range Found on North Atlantic coast from Canada to North Carolina. Naturalized elsewhere. Considered endangered and protected in some states.

Clusters of sea grapes

range Occurs on Atlantic coast from New Jersey to Florida and west to Texas and Oklahoma. Naturalized elsewhere. Propagated commercially and widely used as ornamental.

Wax myrtle

Wax Myrtle

Myrica cerifera
H 6–42' **F** **S**

appearance Simple olive green leaves grow on multiple-branched evergreen shrubs. The female plant produces blue berries with a waxy coating. The leaf blades are fragrant when crushed. Also known as southern bayberry, this plant is used in candle making.

reproduction Depending on the moisture content of its habitat, the wax myrtle reproduces by rhizomes or seeds. In drier areas it tends to form rhizomatous colonies.

habitat This plant grows in sandy soils in a variety of habitats, including coastal sand dunes. Berries form an important fat source for wildlife.

Sea Grape

Coccoloba uvifera
H 6–50' **F** **S**

appearance In dune habitats, sea grape is a low-growing, evergreen shrub with large, round leaves, spikes of whitish flowers, and clusters of reddish, grapelike fruits. As a tree, it can reach as high as 50 feet.

reproduction This plant reproduces by seeds, often dispersed by birds that eat the fruits. Each fruit contains a single oval seed. Flowers on female trees are pollinated, usually by bees, from separate male trees.

habitat Highly salt tolerant, the sea grape is often among the first plant colonizers of sandy and rocky beaches and dunes. It grows well in loamy soils.

range Found along the Florida coast and on the Gulf Coast in Mississippi. Widely naturalized elsewhere.

SOUTH CAROLINA

Ashepoo
Edisto
Combahee

0   10 mi.

Beaufort   St. Helena
Sound

Hunting Island
State Park

Port Royal Sound

Hilton Head Island

Atlantic Ocean

ON LOCATION

Beaufort, South Carolina

- Acres of salt marshes
- Island camping
- Sea turtle nesting
- Large protected areas

The environment of Beaufort, South Carolina, is unique. It is dominated by what look like creeks but are really saltwater arms of Port Royal Sound, snaking through vast marshes of smooth cordgrass (*Spartina alterniflora*). Beaufort County contains half of South Carolina's salt marshes. The area has the highest tidal ranges of any place on the Atlantic coast south of Maine because the curve of the Atlantic coast focuses tides on its center, which is where Beaufort County is located. With a six- to ten-foot range between high and low tides, vast areas of tidal flats are exposed at low tides. When the incoming tide begins covering the

Saltwater marshes thrive in South Carolina's Ashepoo, Combahee, and Edisto River Basin.

flats, fiddler crabs, snails, worms, and other creatures dig in to the pluff mud (the name reportedly comes from the sound your shoes make as you pull them out of the mud). Animals living in the marshes include shellfish, alligators, nutrias, and muskrats. The marshes are also home to shorebirds and wading birds—egrets and blue herons, among others.

Meeting the Marshlands

A good place to begin learning about the complex environments of this part of the coast is Hunting Island State Park on the Atlantic Ocean at the end of U.S. Route 21. It is South Carolina's most popular state park, so visiting on a weekday is probably a good idea. The park offers five miles of natural beach, acres of marsh, tidal creeks, maritime forest, a saltwater lagoon with an ocean inlet and a fishing pier, and a first-rate nature center at the pier. From Memorial Day to Labor Day the park offers public programs that include nature walks and kayak trips.

Hunting Island is a major nesting site for loggerhead sea turtles. During the summer the park offers nighttime beach walks where you might be able to see a female crawling up on the beach to lay eggs or hatchlings breaking out of their eggs and heading for the ocean. Hunting Island has recreational vehicle camping with electrical and water hookups, as well as spaces for tent camping.

Preserving a Unique Environment

Port Royal Sound and its arms, between Beaufort and Hilton Head Island, is an embayment, not an estuary, because no freshwater rivers feed into it. St. Helena Sound on

A great egret in the ACE Basin

the northwestern side of Beaufort County is an estuary fed by the Ashepoo, Combahee, and Edisto Rivers, which give their names to the ACE basin, one of the largest undeveloped estuaries along the Atlantic coast. The basin contains 350,000 acres of land, including towns, farms, and businesses. Almost 40 percent is protected against development, mostly by conservation easements, which bind property owners not to sell or use their land for industrial, commercial, or multifamily development. Owners receive a tax break and preserve a family heritage. This is one of 27 similar U.S. coastal reserves combined into a network of federal, state, and local partnerships that focus on monitoring their environments, research, exchanging information, and education.

Marsh Plants

Marshes, which are pollutant-filtering wetlands created mainly by surface drainage, can be freshwater, brackish, or saline. Most marsh plants emerge on supple stems above their flooded roots. They include sedges, rushes, and grasses—plants with different stem structures and a mnemonic to sort them: "Sedges have edges; rushes are round; grasses are hollow right up from the ground." Marsh plants play a major role in their ecosystem.

Bulrushes blow in a gentle breeze against an azure sky.

Bulrushes

Scirpus species
H 2-10′ **F S**

appearance Bulrushes are sedges with rounded or sharp triangular edges. Some species have inconspicuous, sheathlike leaves, while others have obvious leaf blades. The stem is crowned by small brown spikelets with flowers and seed heads.

reproduction This plant reproduces primarily by rhizomes, especially in shallow water. At greater depths it often produces more seeds.

habitat These species are prolific colonizers of moist soils and shallow freshwater to brackish wetlands, including marshes, lakeshores, riverbanks, and sandbars. The plants form dense stands. Hard-stem species (such as *S. acutus*) grow on a firm substrate. Soft-stem bulrushes (such as *S. validus*) grow in mud.

range Bulrush species found throughout North America, except in far northern areas.

did you know ?

marshes support a variety of bird life, including herons and egrets, which find meals in aquatic nurseries provided by marsh plants? Filter feeders also anchor on plants, which prevent tides from buffeting them.

Smooth cordgrass is an Atlantic coast native.

Smooth Cordgrass

Spartina alterniflora
H 2-7' **S**

appearance Also called low marsh grass, this tall, smooth grass has green stems up to one inch in diameter. It turns brown in fall. The end of the stem is crowned by a long, spiked seed head. Cordgrass withstands saline environments by excreting excess salt through specialized glands in its leaves.

reproduction This plant spreads primarily by extensive underground rhizomes. Less commonly, it spreads by seeds from wind-pollinated flowers.

habitat This species occurs naturally in tidal and salt marshes, where it is the dominant grass species. It is also planted for erosion control along canal banks, levees, dredge fills, and other sites requiring stabilization.

range Native to East Coast of North America from Newfoundland south and along Gulf Coast. Pacific coast populations are non-native.

Common Rush

Juncus effusus
H 2-5' **F S**

appearance Clumps of smooth, round, light green stems rise from sheathed bases. Stems narrow near their ends, where clusters of flowers sprout off the side. The flowers turn yellow and then brown in the fall. Also called soft rush, it is traditionally used in weaving mats and seats. Hundreds of years ago, people also made rush lights by soaking trimmed rushes in fat.

reproduction The rush spreads through short, scaly rhizomes, with new shoots growing in late summer. It also spreads through seeds produced by wind-pollinated flowers and dispersed by wind, water, or animals.

habitat This species grows in fresh and brackish marshes, riverbanks, wet meadows, seeps, pond edges, and ditches. It tolerates polluted conditions and provides food and cover for birds and mammals.

range Found throughout much of North America, except far northern areas. Also commercially propagated and naturalized from cultivars.

Common rushes thrive in salt marshes.

Edge Plants

Cattails waving in a gentle breeze epitomize edge plants. These plants typically establish their roots in water, in a muddy pond bottom or a riverbank, but their main vegetation is above the surface. Others thrive in the moist soil just past the water's edge and tolerate sporadic flooding. Many species do both—grow in and out of the water—and some take advantage of seasonal habitats such as roadside ditches.

Cattails, with their unique flowering spikes

Cattail

Typha species
H 3–10' **F S**

appearance The cattail rises as an erect stem crowned with a velvety spike, accompanied by bladelike leaves. The spike, known as a catkin, is the seed head. As the seeds mature, the catkin becomes downy and the seeds disperse in the wind.

reproduction Catkins produce thousands of seeds. Seedlings spread through aggressive rhizomes, leading to large, monospecific stands of cattails.

habitat Cattails live in fresh to brackish wetlands of all kinds; it thrives in moist soils such as those found at pond edges, swales, and ditches. Traditionally, all parts of the cattail have been put to use for food or fiber.

range Found throughout North America, except in far northern areas. Considered invasive and a threat to biodiversity in many locations.

Pickerelweed

Pontederia cordata
H 2–3' **F**

Pickerelweed

appearance Bright green, heart-shaped leaves emerge on tall stems that also bear spikes covered with tiny bluish-purple flowers. The flowers bloom from the bottom of the spike. The plant filters excess nutrients from the water.

reproduction Rhizomes generate new growth. The plant also reproduces by means of smooth, oval seeds.

habitat Pickerelweed is found in marshes and on pond, lake, and stream edges; it also grows on top of mud. It spreads prolifically and is related to the waterway-clogging water hyacinth. Many waterfowl species feed on the seeds. Its name suggests a tie to the pickerel fish.

range Common throughout eastern North America and in Oregon. Commercially propagated.

The blossom of a blue flag iris

Blue Flag Iris

Iris versicolor
H 2-4' **F**

appearance This showy plant bears flowers with veined, lavender-blue petals and sepals with yellow bases that grow on a sturdy stalk. The leaves are broad and swordlike. The *flag* part of the name comes from the Middle English word for "rush" or "reed."

reproduction This iris spreads through rhizomes and also seed production from flowers. Its yellow sepal bases attract pollinators.

habitat This plant occurs along pond and lakeshores, wet meadows, and other moist areas. It is also widely cultivated.

range Found in northeastern North America as far south as Virginia and west to Manitoba and Minnesota. Also in Idaho.

Skunk cabbage leaves stink when crushed.

Skunk Cabbage

Symplocarpus foetidus
H 1-3' **F**

appearance The plant's yellow-green spadix, covered with tiny flowers, rises first from the ground and is enclosed in a veined, hooded red-brown or purple spathe, or sheath. Later, large, green, tightly rolled, cabbage-like leaves unfold. When leaves are crushed, they give off a fetid stink that attracts flies and other pollinators.

reproduction New plants sprout from underground rhizomes. The species also reproduces from pea-size seeds produced in its flowers. Flowers give off heat that can raise the plant's internal temperature by more than 30°F above that of the air. It may also be wind-pollinated.

habitat Skunk cabbage grows in moist woodlands, swamps, pond edges, and low areas. It often forms a large, carpeted area like a cabbage patch.

range Native in northeastern North America south to South Carolina.

look for **skunk cabbage** very early in the year. Its heat-generating shoots melt through ice and snow, which makes it one of the first plants to emerge.

Pond Plants

Some pond plants live underwater, and their supple stems are unable to reach the surface. Others take root on the bottom and send growth into the air. Still others float with their roots in the water and vegetation on the surface. Pond plants feed and shelter a wide variety of animals, although some prolific species hinder pond biodiversity.

The star-shaped blossom of a water lily contrasts with the plant's deep green pads.

Water Lilies

Nymphaea species
W 2–6" (flower) **F**

appearance This distinctive plant bears a large, green split leaf—the pad—and a star-shaped white or pink blossom that rises on a long stem. Often, each flower blooms for only a few days and then sinks from the surface. Flowers of native species are smaller and less fragrant than introduced tropical ones.

reproduction Water lilies spread by rhizomes and also by seeds, which are a delicacy for ducks. Their flowers are pollinated mainly by beetles and other insects.

habitat Native species are found in quiet waters such as ponds, lakes, and slow-moving streams. Water lilies are widely cultivated commercially and hybridized for use in water gardens.

range *Nymphaea* species found throughout North America, except in far northern areas. Can spread invasively, compromising other vegetation.

did you know ?

native peoples used horsetails, containing abrasive silica, to sand canoes and other wooden items? Colonial settlers, who put them to work on pots and pans, called the useful plants scouring rushes.

A little blue heron wades among water lettuce.

Water Lettuce

Pistia stratiotes
H 3" to 3' 🅵

appearance Thick, ridged, hairy, light green leaves form a rosette like an open, floating head of lettuce, which dangles feathery roots in the water. The rosette can measure up to 18 inches across. Small white flowers top a short stalk and are often hidden by the leaves.

reproduction The plant creates off-shoots supported on short stolons, or stems. It also reproduces through seeds.

habitat Water lettuce is typically found in ponds, lakes, rivers, and canals. Its dense, prolific growth into mats often impedes boating traffic.

range Occurs mainly in southern wet-lands, but appears sporadically as far north as Ohio and New York. Some disagreement as to whether water lettuce is native to U.S. Present since at least 18th century; possibly arrived in ship ballast.

Horsetail

Equisetum hyemale
H 7" to 7' 🅵

appearance An ancient plant dating back about 350 million years, the horse-tail has straight, narrow, segmented, leafless round and hollow stems that form upright clumps. Each segment joint shows a brown or black stripe. The plant often remains green in winter.

reproduction This plant sends out new stems at the segments of its rhi-zomes. It also reproduces by spores that go through a cycle similar to that of ferns. The horsetail's small, brown, pointy cone, called a strobilus, appears only briefly in the spring and dies back after shedding its spores.

habitat The horsetail is found in ponds and marshes, on lakeshores and riverbanks, and in other wetlands. It spreads aggressively and is consid-ered invasive under certain conditions.

range Occurs throughout North America, except in some far northern areas.

Horsetails

A Venus flytrap, a carnivorous plant that catches and digests animal prey

Bog Plants

Bog plants inhabit acidic depressions where dead plant material accumulates and there is no drainage. Low oxygen thwarts decomposition, so nutrient levels stay low and favor adaptive sphagnum moss. Bogs also support plants that satisfy their nitrogen needs by capturing and consuming insects.

Venus Flytrap

Dionaea muscipula
H 4-12" **F**

appearance This plant bears a rosette of specialized hinged and bristled leaves with attractive reddish centers—the traps—and white flowers that rise on leafless stems. The traps clamp shut when an insect visits and brushes the trap's trigger bristles. Enzymes dissolve the insect, thus making nitrogen available to the plant.

reproduction Venus flytraps reproduce mainly by new growth from rhizomes, but also by shiny black seeds produced by flowers. They are widely propagated commercially as a novelty houseplant.

habitat This plant lives in the acidic soils of bogs, seeps, pinelands, and moist, sandy areas on the coastal plain. It has a very localized distribution and is the target of poachers who sell plants. It often perishes due to improper care and feeding.

range Native to southeastern North Carolina and northeastern South Carolina. Also found in New Jersey and Florida. Introduced elsewhere.

safety tip

Bog explorers should tread carefully to avoid breaking through the surface layer, as it is possible to become injured or trapped.

Purple Pitcher Plant

Sarracenia purpurea
H 8-24" **F**

appearance A large, reddish green, veined, hooded, cup-shaped leaf forms the pitcher. Downward-curved hairs trap an insect drawn to the colorful hood. The insect drowns in collected rainwater, and enzymes and bacteria dissolve it. The plant also bears a purplish red pendant flower.

reproduction This pitcher reproduces by rhizomes and by seeds formed in flowers that can be self-pollinated.

habitat This plant lives in peat bogs, swamps, wet grasslands, and some mountain locations. It prefers acidic, sandy soils.

range Extends across Canada, except in far northern areas, to most of eastern U.S. and south to Florida. Also found in California and Washington. Protected in many states. Introduced elsewhere, including Europe.

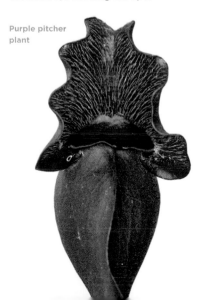

Purple pitcher plant

Sphagnum moss grows in a bog.

Sphagnum Mosses

Sphagnum species
H 4-12" **F**

appearance Tiny leaves in colors of light green, yellow, pink, red, or brown form hairlike tufts close to the stem. The leaf shape varies according to species. Sphagnum has a high water content, holding up to 20 times its own weight in water. Native peoples use it as a diaper. It forms a soft mat on the surface of the bog. The accumulated dead moss underneath is called peat.

reproduction Sphagnum reproduces sexually by creating spores. Spores release in an explosive vortex to aid dispersal, which is important for such a low-growing plant. It also reproduces by leaf fragments. Spores are important for reestablishing populations.

habitat Dozens of sphagnum species are represented in North America. Most are found in bogs, but they can form smaller patches in other acidic wetlands.

range Occurs widely in North America, particularly in northern areas. Extensive commercial production.

Swamp Trees

As denizens of wetlands saturated or flooded by rivers and streams, swamp trees withstand wet "feet" most of the year. They thrive in the dark, nutrient-rich soils, often through structural adaptations that bring needed oxygen to their roots. Unique mangrove species tolerate high salinity in intertidal zones.

Bald cypress trees are draped with Spanish moss in Georgia's Okefenokee Swamp.

Bald Cypress

Taxodium distichum
H 100–150' **F**

appearance This tree is a large, flat-topped conifer with brown or gray ridged bark, widely spaced branches, and featherlike deciduous needles. The trunk spreads at its base. Prominent, cone-shaped "knees" often emerge from submerged roots and help the tree obtain oxygen.

reproduction The tree bears inch-long round cones with angled seeds under the scales, as well as tiny, clustered pollen cones. Its seeds germinate in mud.

habitat The bald cypress lives on wet riverbanks, in swamps, and in other areas with intermittent flooding. It provides nesting sites for bald eagles and ospreys.

range Includes coastal areas south from Delaware and along Gulf Coast, reaching north along Mississippi and Ohio Rivers. Widely propagated and established outside range.

did you **know** ?

a nickname for the bald cypress is wood eternal? The tree boasts decay-resistant heartwood, which makes it ideal for heavy construction, especially for docks and bridges.

Mangroves

various species
H 4–40' **S**

appearance These evergreen species have leathery leaves with glands on the undersides that dissipate excess salt. Adapted structures help take up oxygen: a tangling of above-water arching roots in the red mangrove, or straight spikes, called pneumatophores, that rise from the water in the black mangrove.

reproduction Mangroves reproduce by floating seed dispersal. Seedlings also form on the parent plant, drop off, and float until they can take root in silt.

habitat This tree grows in saline and brackish waters in intertidal zones and mangrove swamps. Some species penetrate inland.

range Found on the southeastern Atlantic coast and Gulf Coast as far as Texas.

Mangroves in the Everglades of Florida

A gnarly water tupelo thrives in a river habitat.

Water Tupelo

Nyssa aquatica
H 50–100' **F S**

appearance This deciduous tree has a tall, straight trunk with a greatly swollen base for stability and an open crown. Long, glossy green leaves turn yellow in the fall. Greenish white flowers appear in the early spring and are pollinated by wind or insects. Dark drupes, or fruits, form in the fall, and each yields a stone with a single seed.

reproduction This tree reproduces from seeds that overwinter and germinate best when buried in mud. Seedlings tolerate some flooding and immersion. It also sprouts prolifically from stumps.

habitat The water tupelo is found in fresh and brackish waters of coastal areas, stream floodplains, and swamps. It often occurs in mixed stands with bald cypress and other species.

range Occurs from Virginia south to northern Florida, west to Texas, and north along the Mississippi Valley.

ON LOCATION

Cape Cod, Massachusetts

- Miles of natural beaches
- Huge dunes instead of condos
- History and nature
- Good hiking and biking

You usually don't have to go far from the surf of an ocean beach to find a variety of environments. A half-mile walk inland from Cape Cod's Race Point Beach to the Province Lands' beech tree forest illustrates this. Tiny sand dunes begin where storms have carried wrack far up the beach. The wrack's nutrients help nourish the dunes' beach grass, which sends out a network of horizontal roots (rhizomes) that help hold the sand together and sends up new grass stalks. Farther inland, you see plants that send roots down to a layer (lens) of rainwater, which soaks through the sand to float on heavier salt water. Insects, spiders, mice, and lizards live here. You might see the tracks of snakes, foxes, and other predators that hunt these smaller animals. Continuing inland, you see more bushes, including blueberries, and then pitch pines, with deep roots that reach water. In a depression that dips into the water lens you might see a cranberry bog roughly six feet across. (Farmers made many larger cranberry bogs.) As you continue inland, the patchy dune plants give way to a forest around ponds in depressions that reach fresh water. Here you find dominant beech trees, tupelo, red maple, inkberry, and swamp azalea. The freshwater ponds' luxuriant life includes arrowhead, pickerelweed, golden club, and water lilies.

Mudflats and Marshes

At low tide on most warm days you will see people exploring the extensive mudflats (also called tidal flats) at the western end of Provincetown Harbor, as well as other exposed Cape Cod flats. The Cape Cod Museum of Natural History in Brewster says its regular Mudflat Mania program for children is its most popular. Out on the flats you can find crabs, shrimp, jellyfish, shellfish, and worms. Some people you see on mudflats are collecting shellfish to eat, as many visitors do. Before going out to pluck

Race Point Lighthouse marks the tip of Cape Cod.

The Stony Brook Grist Mill on Cape Cod sits amid a lush green setting.

dinner from the sand or mud, however, you need to check state regulations for minimum shellfish sizes and buy a permit from the town you will be hunting in.

As almost any coastal area does, Cape Cod offers salt marshes to enjoy. When the ice sheet that covered Cape Cod during the last ice age retreated, it left behind huge blocks of ice that left mostly round holes as they melted. Some became natural cranberry bogs; others filled with fresh water to become kettle ponds or lakes. In the case of the pond in Nauset Marsh, the ocean broke through the barrier, and it is now called Salt Pond. The trail around the pond and marsh includes forests and fields and is a good way to explore a salt marsh ecosystem. At one time the pond was home to ducks and other waterbirds. But in a recent summer, two mute swans were the only birds swimming on the lake; they had chased away other birds, as they do in many habitats.

Exploring Nature on the Cape

The best way to begin a Cape Cod trip is to spend a little time at the National Park Service's Cape Cod National Seashore main visitor's center at Salt Pond just off Route 6 in Eastham. If you have driven to Cape Cod with children, this will be a good stop for restrooms and information. You can learn which ranger-led tours and talks are available, find answers to your questions about Cape Cod National Seashore and other areas, watch short films about the seashore, browse the bookstore, and examine the center's exhibits—including a new one explaining interactions between the Gulf of Maine and Cape Cod. While most visitors come in summer, visiting in spring or fall helps you avoid crowds. You also see animals you won't see in summer, such as migrating birds and harbor seals, which spend summers farther north and arrive at Cape Cod in late fall and winter. You also avoid summer's heat for outdoor activities such as hiking and bicycling.

TOP CHOICES

FROM DR. BEACH

TOP CHOICES FROM DR. BEACH
North America's Finest Water's Edge Locations

"ON LOCATION"

1. Sanibel Island, Florida
2. Calvert Cliffs State Park, Maryland
3. Bay of Fundy, New Brunswick
4. Fitzgerald Marine Reserve, California
5. Big Bend National Park, Texas
6. Port Clinton / Put-in-Bay, Ohio
7. John Pennekamp State Park, Florida
8. Firehole River, Yellowstone National Park
9. Everglades National Park, Florida
10. Great Smoky Mountains National Park, Tennessee & North Carolina
11. Fraser River Estuary, British Columbia
12. Cape May, New Jersey
13. Piedras Blancas Beach, California
14. Barrow, Alaska
15. Beaufort, South Carolina
16. Cape Cod, Massachusetts

ALL-TIME TOP 20 BEACHES

1. Caladesi Island State Park, Dunedin/Clearwater, Florida
2. Coopers Beach, Long Island, New York
3. Fleming Beach Park, Maui, Hawaii
4. Fort De Soto Park, North Beach, St. Petersburg, Florida
5. Grayton Beach State Park, Florida
6. Hanalei Bay, Kaua'i, Hawaii
7. Hanauma Bay, O'ahu, Hawaii
8. Hāpuna Beach State Recreation Area, Big Island, Hawaii
9. Hulopo'e Beach, Lāna'i, Hawaii
10. Kā'anapali, Maui, Hawaii
11. Kailua Beach Park, O'ahu, Hawaii
12. Kapalua Bay Beach, Maui, Hawaii
13. Kaunaoa Beach, Big Island, Hawaii
14. Lanikai Beach, O'ahu, Hawaii
15. Ocracoke Lifeguarded Beach, North Carolina
16. Po'ipū Beach, Kaua'i, Hawaii
17. Siesta Beach, Sarasota, Florida
18. St. Andrews State Park, Florida
19. St. Joseph Peninsula State Park, Florida
20. Wailea Beach, Maui, Hawaii

TOP 10 RIVER SITES

1. Arkansas River, Cañon City, Colorado
2. Colorado River, Glen Canyon, Utah
3. Columbia River, Seaside, Oregon
4. Hudson River, West Point, New York
5. Mississippi River, Hannibal, Missouri
6. Potomac River, Great Falls, Virginia
7. Rio Grande, South Padre Island, Texas
8. Russian River, Cooper Landing, Alaska
9. Sacramento River, Shasta Lake, California
10. St. Lawrence River, Kingston, Ontario

TOP 5 GREAT LAKES BEACHES

1. Bayfield Main Beach, Lake Huron, Ontario
2. Oak Street Beach, Lake Michigan, Chicago, Illinois
3. Presque Isle State Park, Lake Erie, Pennsylvania
4. Sand Point Beach, Pictured Rocks National Lakeshore, Lake Superior, Michigan
5. Sleeping Bear Dunes National Lakeshore, Lake Michigan, Michigan

ALASKA

Pacific Ocean

Hawaiian Islands

For more than 20 years, Stephen Leatherman, known as Dr. Beach, has surveyed coastlines and created an annual list called America's Best Beaches. Here are his all-time choices: the best water's edge destinations in North America.

ALL-TIME TOP 20 BEACHES

Caladesi Island State Park
DUNEDIN/CLEARWATER, FLORIDA

Caladesi Island is only 20 miles west of bustling downtown Tampa. Traveling to this island is an experience in itself, as it is reached only by pedestrian ferry, complete with old salts telling interesting tales, or by private boat. The bay shore is a mangrove forest interlaced with kayak trails and is a bird lover's paradise. The only mode of transportation on this lovely island, which is dominated by palm trees, is by foot over the superfine white sand. There is a great snack bar, as well as other creature comforts to enhance your island experience, and both the grounds and the landscape are well maintained. Lifeguards are on duty during the summer beach season, but the waves are generally only measured in inches, so do not bring your surfboard. On Caladesi the real attractions are the powder white sand and the warm, crystal-clear water, which make it a great beach for enjoyment and relaxation.

Coopers Beach
LONG ISLAND, NEW YORK

Coopers Beach is located in the world-renowned Hamptons on the south shore of Long Island, New York. The water is warm enough for swimming, as it is shielded from the cold Labrador currents that flow off of Montauk Point. As the first Gold Coast in the country, Coopers

Couple on swinging bench, Dunedin, Florida

Beach is hundreds of yards wide and made of grainy white quartz sand. The beach is backed by large sand dunes covered by American beach grass, interspersed with large and extravagant mansions where the rich and famous live and play in the summer. The best way to navigate the small village of Southampton is by bicycle. On your way to the beach, ride through the quiet neighborhoods adorned with huge sycamore trees, privet hedges that are perfectly manicured, and grand estate houses.

D.T. Fleming Park
MAUI, HAWAII

Fleming Park on the western shore of Maui is known for its idyllic, year-round beach weather. This mile-long white-sand crescent beach offers a range of activities and stunning views of nearby Molokai Island. A favorite area for sunbathers, swimmers, and surfers alike, the beach is backed by the welcome shade of a virtual forest of trees on the low sea cliff. Amenities include grills, picnic facilities, and even tablecloth dining on the south end of the park at a nearby four-star hotel. When the surf is up during the winter and spring, stay close to shore unless you are a very strong and experienced swimmer.

Fort De Soto Park, North Beach

Fort De Soto Park, North Beach

ST. PETERSBURG, FLORIDA

North Beach at Fort De Soto Park is a natural jewel on the finger of a sun-drenched city. It is a long, wide, sugar-sand beach with great shelling and lots of nature, including rare birds. There is a 105-year old fort that is a landmark of Florida history. Along with being a wonderful place to swim, the area boasts great fishing, boating, canoeing, kayaking, bird-watching, camping, biking, and trail walking. The beach also offers well-kept facilities and plenty of parking, within this huge park that encompasses five islands.

Grayton Beach State Park

FLORIDA

Grayton Beach is one of a series of great state parks along the Florida Panhandle. The sand is sugar white, the water is emerald green and perfectly clear and clean, and sand dunes dominate the landscape at this state park. Tidal lakes and freshwater ponds punctuate the natural landscape, where only camping is permitted. For those seeking more creature comforts, the old town of Grayton and nearby Seaside offer excellent accommodation and restaurants.

Grayton Beach State Park

Hanalei Bay
KAUAI, HAWAII

Hanalei Bay, on the garden island of Kauai, is a spectacular white crescent-shaped sand beach that many residents and visitors consider to be the most beautiful beach setting in Hawaii. The two-mile-long strand is lined by palm trees, with a backdrop of waterfalls and mountain peaks that reach up to 4,000 feet high. This panorama has been the inspiration for many paintings and photo shoots. The clear emerald waters are perfect for swimming during the tranquil summer months, but only swim at the lifeguarded areas when the surf picks up during the wintertime, as there can be powerful rip currents. The laid-back little village of Hanalei is a good place for refreshments and casual dining, but accommodations are scarce, so plan ahead.

Hanauma Bay
OAHU, HAWAII

Hanauma Bay is a wave-breached volcano, and the crescent-shaped beach is located in the crater. This small, palm-lined beach provides the best snorkeling in

Hanalei Bay

Hanauma Bay

Hapuna Beach

Hawaii in the safe, calm waters protected by an offshore reef. Tropical fish are plentiful in the shallow waters of this marine life conservation area and, along with the spectacular scenery, are the top attraction. This is the first beach in the nation to be smokeless in order to eliminate the problems of cigarette butts on the beach and in the crystal-clear waters. It is necessary to come early, as the parking lots fill up quickly.

Hapuna Beach State Recreation Area
BIG ISLAND, HAWAII

Hapuna is a white coral sand beach that contrasts with the black lava rock that flanks and bounds this half-mile pocket beach. The emerald green waters are great for swimming, snorkeling, and scuba diving during the summer months when there is little to no surf. During big wave days in the winter, it is best to stay on the beach or to play only at the edge of the water, because powerful shore breaks and rip currents make swimming impossible. The weather is hot, dry, and sunny throughout most of the year, which makes Hapuna an ideal playground in the winter, but some may find the heat a bit overwhelming in the summer.

Hulopoe Beach
LANAI, HAWAII

Lanai is one of the least visited islands in Hawaii, and it has only one great swimming beach—but it is a real gem. Hulopoe Beach is a lovely white-sand beach lined with palm trees. This long, crescent-shaped beach is anchored by a lava flow to the east and bounded to the west by a high sea cliff where a luxurious hotel is located. The crystal-clear waters make for great swimming except during a Kona storm, when heavy swell waves roll in from the south. Hulopoe is an idyllic place where native Hawaiians and visitors alike can enjoy this getaway beach.

Kaanapali
MAUI, HAWAII

Kaanapali is the best known beach in Maui, with a string of hotels along its four-mile-long strand of white coral sand. As in all Maui resort developments, the hotels at Kaanapali are set back nicely from the water, which provides ample space for a beach promenade and tropical flowers and coconut trees. The beach is divided into two sections by a point of lava rocks that creates variations in wave conditions along this long beach, so that some areas have lower surf and better wading conditions for children. The island of Lanai is just offshore, and humpback whales arrive in large numbers during the winter to breed in these waters.

Kailua Beach Park
OAHU, HAWAII

Kailua is little more than a half hour's car drive from Honolulu, but crossing the mountain range takes you to a relaxed beach that seems a world away. Here on the windward shore of Oahu is one of the best places to set sail in a small boat or to learn to windsurf, as the wind is always onshore. This 30-acre beach park has a good amount of parking, shade trees, and a fine sandy beach. The beach gradually slopes offshore, and because there are no hazardous water conditions, it is a very safe area for swimming. The village of Kailua, with its many restaurants and shops, is within walking distance.

Kapalua Bay Beach

MAUI, HAWAII

Travelers often rate Maui as the best resort island in the world, and the little pocket beach at Kapalua is my personal favorite. The name Kapalua means "arms embracing the sea," because lava flows bound this beautiful crescent beach and keep it safe from dangerous waves and currents. The tranquil, clear waters are very inviting for swimming, and just offshore are living corals that attract schools of multicolored tropical fish. You can buy fish food and rent snorkeling gear at the concession hut on one end of the beach, and dine at a great restaurant on the rock that anchors the other end of the beach. Public parking is provided at this exclusive resort, which was tastefully developed on a former pineapple plantation.

Kailua Beach Park

Ocracoke Lifeguarded Beach

Kaunaoa Beach

BIG ISLAND, HAWAII

Kaunaoa is an oasis by the sea in an otherwise bleak landscape dominated by barren black lava rock. This long white-sand beach, bounded by lava flows jutting into the sea, is a great spot to swim and to spot many sea creatures, such as the rare Hawaiian green sea turtles, in the sparkling clear waters. The southern end of the beach is the most protected and offers a wide, flat place for children to play. Kaunaoa, which is also known as Mauna Kea Beach because of the luxurious hotel there, is a public beach, so just let the gatekeeper know that you are visiting this beach.

Lanikai Beach

OAHU, HAWAII

Lanikai, with its clear, emerald green waters and two offshore islands, is one of the most picturesque beaches in Hawaii. People are invariably intrigued by the water coloration, caused by the rare combination of white, highly reflective sand and very clean, clear, shallow water through which the sun shines and is also reflected. The shallow water is excellent for swimming, and there is good snorkeling further offshore around the coral heads. The beach is shifting in position, so there is little to no beach in some areas, while the sand builds up farther down the shore. There is no public parking area, but all Hawaiian beaches are public, so park along the street in this shore enclave and walk to the end of the street to gain access.

Ocracoke Lifeguarded Beach
NORTH CAROLINA
Ocracoke Island on the Outer Banks of North Carolina is like traveling back in time—you can really get away from the hustle and bustle on this remote island. This is not the place you go to enjoy a Hilton spa or to play a round of golf. Incredibly, the 40-minute car ferry ride is still free, and then you find yourself on 14 miles of wild and empty beaches, with a miniature village on the south end. Bikes are the transport of choice on the island, which was once the stomping grounds of Blackbeard the pirate—as well as the place where he met his untimely death. For families with small children, the best time to visit the Outer Banks is in June or July, since late-summer hurricanes can create strong waves and rip currents.

Poipu Beach
KAUAI, HAWAII
Poipu Beach Park is one of the most popular swimming areas in Kauai and is frequented by residents and tourists alike. Offshore rocks provide more protected waters at one flank of the beach, while a popular tourist hotel is located on the other end of the beach, where surfers and strong swimmers go. The coral sand has a golden hue, making it especially beautiful at sunset. This is the dry side of Kauai, which means that the sun shines brightly while the beautiful tropical flowers on the rest of Kauai are being naturally watered.

Siesta Beach
SARASOTA, FLORIDA

Siesta Beach boasts the finest, whitest sand in the world, which attracts sand collectors from far and wide. Its clear, warm waters are ideal for swimming and bathing. The beach is hundreds of yards wide and crescent shaped, due to anchoring of onshore rocks to the north and a unique underwater formation of coral rock and caves. This provides great conditions for snorkeling and scuba diving. Siesta is great for volleyball and other types of recreational fitness. The small village of Siesta Beach is within walking distance. The southwest coast of Florida is a favorite of snowbirds from the Midwest, so it is really a year-round beach. My favorite time to visit is late fall, as the water is still quite warm in November—although I have been swimming here as early as February.

St. Andrews State Park
FLORIDA

St. Andrews is a beach that is described in superlatives for nature lovers—clear, emerald waters, pearly white sand, and a naturally sculpted landscape located near the amenities of Panama City Beach. Fishing from the inlet jetty is super, as the tidal currents carry many game fish past your lure. Across the inlet is Shell Island, with seven miles of undisturbed beach that is, as the name implies, a great place to beachcomb for shells. Birding is also a popular activity here, as well as swimming in the warm Gulf waters.

St. Joseph Peninsula State Park
FLORIDA

This huge, arc-shaped barrier beach is reachable from the mainland across Apalachicola Bay and via Cape San Blas. The estuarine region is the largest producer of seafood in Florida, and Apalachicola oysters are renowned for their flavor. The fine, snow-white sand forms fantastic beaches for swimming and walking, and onshore winds have produced large barrier dunes. These sand dunes, which are 30 to 40 feet high, are spanned by wooden walkovers to provide passage from the camping areas to the beach. This 2,500-acre beach park provides a lot of nature, peace, and quiet. There is a limited number of cabins in the park; commercial accommodations are available at the park's entrance. A great time of year to visit is during the fall, when the water is still warm.

Wailea Beach
MAUI, HAWAII

Wailea is one of Hawaii's most luxurious beach resort destinations; it is literally an oasis of greenery in a semi-arid climate. A series of five pocket beaches bounded by black lava points provide great swimming conditions in the calm waters. All the white sandy beaches are connected by a coastal walk and a landscaped minipark that stretches for two miles. Ulua Beach, the middle beach at Wailea, has the clearest and best waters for snorkeling. The sand is reddish in color, as it is a mixture of coral sand and lava-derived sediments. The rich and famous frequent Wailea's elegant beachfront hotels, but public parking and beach access are provided when requested.

Siesta Beach, Sarasota

Wailea Beach

Colorado River, Glen Canyon, Utah

TOP 10 RIVER SITES

Arkansas River

SALIDA, COLORADO

The Arkansas River is a spectacular area for viewing arid scenery while taking an exciting white-water raft trip. Many of the trips start out at Salida or Cotopaxi and run downstream toward Cañon City. The river is safe for families with Class III rapids for great excitement.

Colorado River

MOAB, UTAH

Everyone is familiar with the world-famous Grand Canyon and Hoover Dam, but another place to experience the Colorado River is Moab. As an undergraduate student I had the experience of conducting geologic mapping at nearby Arches National Park, and our bath was a dip into the Colorado River. The temperature dropped from 110 (air) to around 50 (water). The best way to experience the river is by rafting; trips range from relaxing to adrenaline rushes.

Columbia River

SEASIDE, OREGON

The Columbia River is one of America's mighty rivers, and nowhere is it more exciting to see wave action than where this river meets the Pacific Ocean. The U.S. Coast Guard uses the Columbia River bar as a test for new recruits; sometimes these 42-foot cutters do a 360° roll (with the sailors roped into their seats) when hit broadside by a huge breaking wave. Seaside—a small, artsy beach town nearby—is certainly worth a visit.

Hudson River

WEST POINT, NEW YORK

Most people experience the Hudson River when crossing it by bridge from New Jersey to Manhattan. My favorite place to view the river is at the U.S. Military Academy at West Point. West Point was chosen for the USMA because the river makes a big bend here, providing a military advantage during the Revolutionary War to the Patriots and great viewing today. I also enjoy watching the cadets marching across the broad fields in full uniforms.

Mississippi River

HANNIBAL, MISSOURI

New Orleans is the most famous city on the Mississippi, but a visit to Hannibal, boyhood home of Mark Twain, is a walk back in time. The Mississippi is a very meandering river, unlike many other river channels, and it changes dramatically along the more than 2,000 miles of its course, from Lake Itasca, Minnesota, to its delta in Louisiana.

Potomac River

GREAT FALLS, VIRGINIA

The Potomac River transitions from fast-moving to sluggish at Great Falls, which is the boundary between the hard-rock piedmont and the unconsolidated sediments of the much flatter coastal plain. This north-south "fall line" marks the furthest landward extent of sea level in geologic times. Great Falls Park is a splendid place to have a picnic lunch and watch the kayakers try their luck negotiating the white-water rapids.

Rio Grande
SOUTH PADRE ISLAND, TEXAS

The Rio Grande forms the boundary between Mexico and the United States until it enters the Gulf of Mexico. South Padre Island has the clearest water on the Texas coast and is great for swimming. Many wildlife and citizen snowbirds from the North American Midwest overwinter here. Nearby Port Isabel is known for its fishing charters.

Russian River
COOPER LANDING, ALASKA

The Russian River Falls overlook area is a great place to see grizzly bears catching and eating salmon. The Russian River flows into the Kenai River, where a peaceful white-water raft trip affords views of many eagles and people lining the bank to fish during the summer salmon runs.

Rio Grande

Russian River

Sacramento River

SHASTA LAKE, CALIFORNIA

This 400-mile-long river, largest in California, originates in the Klamath Mountains and empties into San Francisco Bay. Many of its miles have been altered to supply water, especially for agriculture in the Sacramento Valley. Shasta Lake, for example, was formed by building a massive concrete dam, tallest in the U.S., between 1935 and 1945. To the north, the river offers superb springtime white-water rafting with majestic views of Mount Shasta.

St. Lawrence River

KINGSTON, ONTARIO

Kingston, location of Fort Henry, garrison during the War of 1812, is a good starting point from which to see Lake Ontario become the St. Lawrence River, a mighty commercial and natural waterway. Thirty miles downriver, visit the Thousand Islands International Bridge, a 4,500-foot suspension bridge connecting New York and Ontario passing through the beautiful Wellesley Island State Park.

Sacramento River

St. Lawrence River

Sleeping Bear Dunes National Lakeshore

TOP 5 GREAT LAKES BEACHES

Sleeping Bear Dunes National Lakeshore

LAKE MICHIGAN, MICHIGAN

Here you will find some of the most spectacular coastal dunes in the world. I especially liked trekking up the huge dune on the north end of this expansive park. There is a variety of beaches with fine sand and crystal-clear water. Children often look for Petoskey stones (beautiful pebbles of fossilized coral) in the small, shallow creeks that drain into Lake Michigan and have warmer water than the lake's. Traverse City is a great little nearby town with a lot of attractions.

Presque Isle State Park

LAKE ERIE, PENNSYLVANIA

This seven-mile-long, naturally occurring sand spit provides a range of water-based activities, ranging from 13 principal swimming areas (which draw some crowds) to unique ecological areas with very few people. There is so much to do and see in this curving peninsula-like feature that juts into Lake Erie that it is certainly worth the trip to the 3,200-acre park.

Sand Point Beach,
Pictured Rocks National Lakeshore
LAKE SUPERIOR, MICHIGAN
Offshore islands protect this small beach from waves, and the water is shallow. These conditions make the water warm enough to go swimming, which I did in August. This is a huge park with spectacular and picturesque cliffs, which are best seen on a boat trip. The water is crystal clear and surprisingly green colored in shallow areas.

Bayfield Main Beach
LAKE HURON, ONTARIO, CANADA
The small village of Bayfield on Lake Huron boasts of being a Blue Flag beach, signifying its environmental quality and services. There are amenities close to the beach, such as bathrooms, and plenty of parking as crowds are never a problem at this remote beach. The beach is sandy, but grades into cobbles offshore in many areas. Bayfield is a delightful town to explore, and the town pier is the favorite place to watch sunsets.

Oak Street Beach, Lake Michigan

Oak Street Beach
LAKE MICHIGAN, CHICAGO, ILLINOIS
Oak Street Beach is located at the north end of Chicago's famed Lake Shore Drive. Cyclists, runners, and parents with strollers use this 18-mile lakefront; the fine-sand beach has lifeguards and high water quality that makes swimming a pleasure. This famous beach, which dates to the 1890s, has stunning views of the historic architecture of Chicago as well as Lake Michigan. I recommend taking public transportation to the beach, but there is some street parking nearby.

FURTHER RESOURCES

Safety at the Water's Edge

When exploring the water's edge, it is important to keep safety in mind. According to the Centers for Disease Control, more than a thousand people drown in natural water settings in the United States each year. Preparation can help prevent injuries and drowning.

General Safety Tips
- Learn to swim.
- Supervise children closely. Make sure they wear approved personal flotation devices when in or around the water.
- Never swim alone.
- Whenever possible, swim in areas supervised by lifeguards.
- Both children and adults should wear a regulation life jacket when in a boat.
- Learn CPR to be prepared for emergencies.
- Stop swimming at the first sign of bad weather.
- Enter the water feetfirst, never headfirst, unless the area is designated for diving.
- Never mix alcohol with swimming, diving, or boating.
- Drink plenty of water regularly to keep your body hydrated.

Safety at the Beach
- Be aware of a beach's potential hazards, such as deep and shallow areas, currents, depth changes, and obstructions.
- Never turn your back to the ocean. You can be swept off by waves that can come without warning.
- Swim parallel to shore if you want to swim long distances.
- Wear sunglasses to protect your eyes from damage that can occur from UV rays.
- Wear foot protection. Feet can get burned, especially on black sand.
- Prevent sunburns by using sunscreen with a protection factor (SPF) of at least 30.

Safety in Rivers and Lakes
- Be sure rafts and docks are in good condition.
- Never swim under a raft or dock.
- Always look before jumping off a dock or raft to be sure no one is in the way.
- Teach children that if they lose sight of their group, they should sit down and stay put.
- Give everyone a police whistle. If anyone becomes lost, he or she should blow the whistle three times, wait a little, and repeat.

Dangerous Waves and Currents
- A rip current is a powerful channel of water that flows from the beach out into the ocean. It can drag swimmers away from the beach. The United States Lifesaving Association estimates that rip currents kill an average of 100 people each year on U.S. beaches and account for more than 80 percent of beach rescues each year. Don't try to fight the current to get back to shore. Instead, swim parallel to the beach to escape the relatively narrow rip current.
- Longshore currents are beach currents that move parallel to the shore. A strong longshore current can sweep swimmers into hazardous areas and sometimes make it difficult to return to the beach.
- Shore breaks—tall waves that break directly onto the face of the

beach—are the biggest wave danger beachgoers face. Shore breaks usually occur on steep beaches, where the bottom drops quickly. A particularly strong shore break can slam a person onto the beach.

- Surf beats are groups in which waves often travel. If you find yourself face-to-face with a big breaking wave, take a deep breath, dive under it, and come up on the other side. Do not try to jump it, because it may lift and throw you.
- Backwash is the return flow of water from a broken wave as it heads back to the ocean. When waves get tall and powerful, the backwash can be swift, deep, and forceful enough to knock your feet out from under you.
- Rogue or sleeper waves—upstart waves that do not occur in a general pattern of large waves—are large and spontaneous. They are uncommon on sandy beaches, however.

Weather Hazards

A lightning strike is the most serious weather-related hazard when you are in or near the water. Water is a good conductor of electricity, so any water sport—such as swimming, snorkeling, surfing, or Jet Skiing—is hazardous when lightning is near. The greatest danger comes at the beginning and end of a thunderstorm, when people are least likely to be under cover. Lightning can strike from ten miles away, so if you can hear it, it is close enough to be a hazard.

- Listen to weather reports before going to the beach.
- Watch for dark clouds approaching, especially those that come from inland in the afternoon.
- At the sound of thunder or the sight of lightning, get out of the water right away and find sturdy shelter.

- Avoid being near trees or other tall objects during a lightning storm.
- A group of people should spread out so that if lightning hits one, the others can begin CPR and call or run for help.
- Wait at least 5 minutes—and preferably 30—before returning to the water after a storm.

Flash floods, which bring quickly rising water, are another weather hazard. A potential danger along any stream or river, these floods are especially dangerous in the dry West. Thunderstorms there are often too far away for people to hear thunder or to see the clouds, which can send deep water roaring down a dry or nearly dry streambed with little warning. In the East, flash floods usually occur during or soon after widespread heavy rain. Check weather forecasts before leaving for a watery exploration. Then watch the sky for any dramatic changes.

If a hurricane is forecast, pay attention to weather reports. There is no reason to be on an ocean beach or estuary when a tropical storm or hurricane hits, since forecasters in North America give a heads-up five or more days in advance and issue specific alerts at least two or three days before a storm hits. And remember, a hurricane does not have to hit a coast to be deadly. In 2009, Hurricane Bill's center never came closer than 160 miles from the U.S. East Coast, yet its waves killed a man in the surf at New Smyrna Beach, Florida, as well as a girl who was watching from rocks as waves pounded the shore at Maine's Acadia National Park.

Animal Dangers

More deaths are attributed to bee stings than to shark attacks; even so, you should not ignore the small dangers that sharks and other marine animals present. Watery worlds such as oceans are wild territories, not aquariums or zoos with barriers between you and the animals. As such, unexpected encounters between swimmers and sea life can occur. Knowledge and preparedness are the best protection.

Most shark attacks occur in warm water, and about 90 percent of them happen at or near the surface of the ocean. Three shark species are responsible for nearly all the life-threatening attacks on humans: the great white shark, the tiger shark, and the bull shark. Even though the likelihood of a shark attack is low, here are precautions to make one even less likely:

- Swim in a group.
- Don't swim with an open wound.
- Don't swim at twilight or at night.
- Don't swim with pets.
- Avoid dirty or murky water.
- Don't wander too far from shore.
- If you spot a shark, swim smoothly and exit the water.
- Avoid thrashing and splashing movements.
- Avoid shiny jewelry.
- Be careful when swimming near steep underwater drop-offs, jetties, or piers.

You should also be careful to avoid jellyfish stings. Many jellyfish, but not all, have tentacles that can slide across or wrap around arms or legs without the swimmer even seeing the creature. Although few such stings are debilitating, they are painful. If someone gets stung, rinse the affected area with seawater. Remove any tentacles from the skin, and keep an eye on the victim. If signs of an allergic reaction occur, call 911 immediately.

The Portuguese man-of-war (not a jellyfish but a hydroid) inflicts the worst sting. Iridescent and purple, these floating creatures have long tentacles that can grow to 50 feet in length. High winds and storm conditions can blow Portuguese man-of-wars onto beaches, where they sting even as they lie motionless. As with jellyfish stings, rinse the area with seawater and pick off visible tentacles. Apply ice for pain and seek medical attention if the victim appears to be in shock.

Stingrays, which sometimes hover in shallow water, do not attack people. However, if a swimmer inadvertently steps on one, the animal may defend itself by slapping its spurred tail at the foot or leg. To avoid this, do the stingray shuffle when you wade in the water: Shuffle along instead of taking steps. This gives the sea critter a warning of your presence and a chance to escape.

Other animal life to watch out for:
- Bluefish, which are a predatory fish and have extremely sharp teeth. These aggressive fish can bite people in the midst of a feeding frenzy.
- Sea urchins, which live in tidal pools attached to rocks or coral. If a beachcomber steps on one, the spines penetrate the foot and cause pain and possibly infection.
- Ticks and chiggers, which can be found near watery environments. Ticks are the more dangerous of the two since they can spread Lyme disease or Rocky Mountain spotted fever. Chiggers are not known to carry any human diseases, but they cause painful itching. Protect

yourself by wearing long pants tucked in to socks.

- Giardia protozoa, which cause an intestinal disease and are often found in mountain streams. These critters are not discernable to the naked eye, so to be safe, do not drink from mountain streams that appear clean. The water is rarely as pure as it looks.

Plant Dangers

- Keep away from poison ivy (and its relatives, poison oak and poison sumac). Poison ivy secretes an oily, clear liquid—urushiol—that can cause a painful, itchy allergic rash. Its favored habitat across much of North America is on the edge of wooded areas, such as along shores.
- Do not eat any part of any plant unless you are absolutely sure what it is, even if it looks like something familiar. For example, the berries of the deadly night-shade plant look somewhat like cherry tomatoes.
- If you are interested in collecting wild mushrooms, do your homework and learn how to tell the safe ones from the killers—even if the mush-rooms you see seem to be like those you buy in the market.

Pollution Safety

In most public beaches and along the water's edges, precautions are taken to protect visitors from pollution and the problems it can bring. Even so, it is important to be informed so you can protect yourself.

- In locations where possible runoff contamination is high, wait two to three days after a heavy rain before swimming in the water.

- Always swim more than 100 feet away from storm drains that flow into the sea.
- Always swim updrift of the current's flow.
- Avoid swimming near marinas, where bacterial levels in the water can be high.
- Never play in runoff ponds or in storm drain water at the water's edge.

Also, find out if red tide occurs at the site you plan to visit. Red tide is a condition produced by a bloom of algae in the water. It is not always red—it can be green or brown—and is signaled by masses of organic particles floating on top of the water. While red tides are a natural phenomenon, some scientists believe that industrial and agricultural runoff is exacerbating the problem. No one should swim at a place where such a bloom is detected. The toxins in red tide can not only make swimmers ill, but also sicken people who eat shellfish that grows in these waters.

FURTHER READING

General

Ackerman, Jennifer. *Notes from the Shore.* Penguin, 1996.

Caduto, Michael J. *Pond and Brook: A Guide to Nature in Freshwater Environments.* UPNE, 1990.

Gibbons, Euell. *Stalking the Blue-eyed Scallop.* David McKay, 1964.

Leatherman, Stephen, and John Fletemeyer, eds. *Rip Currents: Beach Safety, Physical Oceanography, and Wave Modeling.* CRC Press, 2011.

Russell, Franklin. *Watchers at the Pond.* Nonpareil Books, 2000.

Trefil, James S. *A Scientist at the Seashore.* Dover Books, 2005.

Weiss, Judith S. *Salt Marshes: A Natural and Unnatural History.* Rutgers University, 2009.

Chapter 1—Beachcombing

Gosner, Kenneth L., and Roger Tory Peterson. *Peterson Field Guide to the Atlantic Seashore: From the Bay of Fundy to Cape Hatteras.* Houghton Mifflin, 1999.

Iselin, Josie. *Beach: A Book of Treasures.* Chronicle, 2010.

Morris, Percy A., R. Tucker Abbott, and Roger Tory Peterson. *Peterson Field Guide to Shells of the Atlantic and Gulf Coast and the West Indies,* 4th ed. Houghton Mifflin, 1995.

Morris, Violet F., and Roger Tory Peterson. *Peterson Field Guide to Pacific Coast Shells,* 2nd ed. Houghton Mifflin, 1974.

Rehder, Harald A. *National Audubon Society Pocket Guide to Familiar Seashells.* Knopf, 1988.

Rothschild, Susan B., and Nick Fotheringham. *Beachcomber's Guide to Gulf Coast Marine Life,* 3rd ed. Taylor Trade, 2004.

Sept, J. Duane. *The Beachcomber's Guide to Seashore Life of California,* rev. ed. Harbour, 2009.

Shumway, Scott W. *The Naturalist's Guide to the Atlantic Seashore.* Falcon, 2008.

Witherington, Blair. *Florida's Living Beaches: A Guide for the Curious Beachcomber.* Pineapple Press, 2007.

Chapter 2—Invertebrates

Gowell, Elizabeth. *Amazing Jellies: Jewels of the Sea.* Bunker Hill, 2004.

Lamb, Andy, and Bernard Hanby. *Marine Life of the Pacific Northwest: A Photographic Encyclopedia of Invertebrates, Seaweeds and Selected Fishes.* Harbour, 2005.

Meinkoth, Norman A. *The Audubon Society Field Guide to North American Seashore Creatures.* Knopf, 1981.

Sprung, Julian. *Corals: A Quick Reference Guide.* Ricordea, 1999.

Chapter 3—Insects & Other Invertebrates

Borror, Donald J., and Richard E. White. *Peterson Field Guide to Insects: America North of Mexico.* Houghton Mifflin, 1998.

Brock, Jim P., and Kenn Kaufman. *Kaufman Field Guide to Butterflies of North America.* Houghton Mifflin, 2006.

Dunkle, Sidney W. *Dragonflies Through Binoculars: A Field Guide to Dragonflies of North America.* Oxford University Press, 2000.

Eaton, Eric R., and Kenn Kaufman. *Kaufman Field Guide to the Insects of North America.* Houghton Mifflin, 2007.

Evans, Arthur V. *Field Guide to Insects and Spiders of North America.* Sterling, 2007.

Paulson, Dennis. *Dragonflies and Damselflies of the West.* Princeton University Press, 2009.

Voshell Jr., J. Reese. *A Guide to Common Freshwater Invertebrates of North America.* McDonald & Woodward, 2002.

Waldbauer, Gilbert. *A Walk Around the Pond: Insects in and over the Water.* Harvard University Press, 2008.

Chapter 4—Fish

Gilbert, Carter R. *National Audubon Society Field Guide to North American Fishes.* Knopf, 2002.

Hoese, H. Dixon. *Fishes of the Gulf of Mexico: Texas, Louisiana, and Adjacent Waters,* 2nd ed. TAMU Press, 1998.

Lamb, Andy, and Phil Edgell. *Coastal Fishes of the Pacific Northwest,* 2nd ed. Harbour, 2010.

Page, Lawrence M., and Brooks M. Burr. *Peterson Field Guide to Freshwater Fishes,* 2nd ed. Houghton Mifflin, 2011.

Ray, Carleton, and C. Richard Robins. *Peterson Guide to Atlantic Coast Fishes.* Houghton Mifflin, 1999.

Chapter 5—Reptiles & Amphibians

Conant, Richard, and Joseph T. Collins. *Peterson Guide to Reptiles and Amphibians of Eastern and Central North America,* 4th ed. Houghton Mifflin, 1998.

Elliot, Lang. *The Frogs and Toads of North America.* Mariner Books, 2009.

Spotila, James R. *Sea Turtles: A Complete Guide to Their Biology, Behavior, and Conservation.* Johns Hopkins, 2004.

Stebbins, Robert C. *Peterson Guide to Western Reptiles and Amphibians,* 3rd ed. Houghton Mifflin, 2003.

Chapter 6—Birds

Chandler, Richard. *Shorebirds of North America, Europe, and Asia.* Princeton University Press, 2009.

Dunn, Jon, and Jonathan Alderfer. *National Geographic Field Guide to the Birds of North America,* 6th ed. National Geographic, 2011.

O'Brien, Michael, et al. *The Shorebird Guide.* Houghton Mifflin, 2006.

Paulson, Dennis. *Shorebirds of the Pacific Northwest.* University of Washington Press, 1993.

Chapter 7—Mammals

Bowers, Nora, and Rick and Kenn Kaufman. *Kaufman Field Guide to Mammals of North America.* Houghton Mifflin, 2004.

Murie, Olaus J., and Mark Elbroch. *Peterson Field Guide to Animal Tracks,* 3rd ed. Houghton Mifflin, 2005.

Reid, Fiona. *Peterson Field Guide to Mammals of North America.* Houghton Mifflin, 2006.

Rezendes, Paul. *Tracking and the Art of Seeing: How to Read Animal Tracks and Signs,* rev. ed. HarperCollins, 1999.

Stewart, Brent S., Phillip J. Clapham, James A. Powell, and Randall R. Reeves. *National Audubon Society Guide to Marine Mammals of the World.* Knopf, 2002.

Chapter 8—Plants

Mondragon, Jennifer. *Seaweeds of the Pacific Coast: Common Marine Algae from Alaska to Baja California.* Sea Challengers, 2003.

National Audubon Society. *Field Guide to the Wildflowers of North America: Eastern Region,* rev. ed. Knopf, 2001.

Rushforth, Keith, and Charles Hollis. *Field Guide to the Trees of North America.* National Geographic Society, 2006.

Sibley, David. *The Sibley Guide to Trees.* Knopf, 2009.

Thomas, David. *Seaweeds.* Smithsonian Books, 2002.

INDEX

Bold page numbers indicate illustrations.

ILLUSTRATIONS CREDITS

Cover: Ian Cumming/Getty Images; low left, Medford Taylor/ National Geographic Stock; center left, Darlyne A. Muraski/ National Geographic Stock; center right, Tay Tousey, National Geographic My Shot; low right, Flip de Nooyer/Foto Natura/ Minden Pictures/National Geographic Stock; back cover: low left, Whit Richardson/Getty Images; center left, Greg Vaughn/ Getty Images; center right, Frederic Pacorel/Getty Images; low right, Ralph Lee Hopkins/National Geographic Stock; spine, Richard A. Cooke/Corbis.

1, Alberto Pomares/Getty Images; 2-3, Monica and Michael Sweet/Getty Images; 4, Alberto Pomares/Getty Images; 11, Skip Brown/National Geographic Stock; 12 (LE), Alaska Stock Images/National Geographic Stock; 12 (RT), Steve Cole/Getty Images; 13, Jupiterimages/Getty Images; 14, James Forte/ National Geographic Stock; 16 (LE), Rachid Dahnoun/Aurora Photos; 16 (RT), Jereme Thaxton/Aurora Photos; 19 (UP LE), Jim McKinley/Getty Images; 19 (UP RT), Crawford A. Wilson III/ Getty Images; 19 (LO LE), Datacraft Co Ltd./Getty Images; 19 (LO RT), Greg Kessler/Getty Images; 20, Stephen Leatherman; 21, David Patrick Valera/Getty Images; 23 (UP), Gordon Wiltsie/ National Geographic Stock; 23 (LO), Matt Champlin/Getty Images; 24, Skip Brown/National Geographic Stock; 25 (LE), Darlyne A. Murawski/Getty Images; 25 (UP RT), Carr Clifton/ Minden Pictures/National Geographic Stock; 25 (LO RT), Jim Brandenburg/National Geographic Stock; 26 (LE), Ocean/ Corbis; 26 (CTR), Paul Edmondson/Getty Images; 26 (RT), Dale Wilson/Getty Images; 27 (LE), Cameron Davidson/Getty Images; 27 (CTR), Ron Chapple/Getty Images; 27 (RT), Eric Full/ Shutterstock; 29, Ralph Lee Hopkins/National Geographic Stock; 30, USGS/NASA; 33, Michael Melford/National Geographic Stock; 35, PPSOP/Corbis; 36, Raymond Gehman/ National Geographic Stock; 37, Mike Theiss/National Geographic Stock; 40, Gary Hincks; 41, Raymond Gehman/National Geographic Stock; 43, Mark Raycroft/Minden Pictures/National Geographic Stock; 44 (UP), Kevin Schafer/Getty Images; 44 (LO), Brian Gordon Green/National Geographic Stock; 47 (UP LE), Ed Reschke/Getty Images; 47 (UP CTR), Frans Lanting// National Geographic Stock; 47 (UP RT), Raymond Gehman/ National Geographic Stock; 47 (CTR LE), Robbie George/ National Geographic Stock; 47 (CTR), Karen Kasmauski/ National Geographic Stock; 47 (CTR RT), Gerry Ellie/Minden Pictures/National Geographic Stock; 47 (LO LE), Michael Melford/National Geographic Stock; 47 (LO CTR), Stephen Sharnoff/National Geographic Stock; 47 (LO RT), Cisca Castelijns/ Foto Natura/Minden Pictures/National Geographic Stock; 48 (UP LE), Robert Yin/Corbis; 48 (UP CTR), Robert Marien/Corbis; 48 (UP RT), Norbert Wu/Corbis; 48 (LO LE), Paul Souders/ Corbis; 48 (LO CTR), Stephen Frink/Corbis; 48 (LO RT), David Wrobel/Corbis; 49, Jun Zuo/National Geographic My Shot; 50, George Grall; 51 (LE), Frans Lanting/National Geographic Stock; 51 (RT), Stephen St. John/National Geographic Stock; 53, Annie Griffiths/National Geographic Stock; 56, George Grall/National Geographic Stock; 58, Medford Taylor/National Geographic Stock; 59 (UP), Ivonne Wierink/Shutterstock; 59 (LO), David Wilkins/Shutterstock; 60 (UP), B. Anthony Stewart/National Geographic Stock; 60 (LO), Raul Touzon/National Geographic Stock; 61 (UP), James Randklev/Getty Images; 61 (LO), DK Limited/Corbis; 62 (LE), Darlyne A. Murawski/National Geographic Stock; 62 (RT), Robert Clark/National Geographic Stock; 63 (UP), Medford Taylor/National Geographic Stock; 63 (LO), Victor R. Boswell, Jr./National Geographic Stock; 64, W. Treat Davidson/National Geographic Stock; 65 (UP), Visuals Unlimited, Inc./Robert & Jean Pollock/Getty Images; 65 (LO), Victor R. Boswell, Jr./National Geographic Stock; 66 (UP), Victor R. Boswell, Jr./National Geographic Stock; 66 (LO), Visuals Unlimited, Inc./Robert & Jean Pollock/Getty Images; 67 (UP), Phillippe Clement/naturepl.com; 67 (CTR), Wikipedia; 67 (LO), Jurgen Freund/naturepl.com; 68 (UP), Wolcott Henry/National Geographic Stock; 68 (LO), Victor R. Boswell, Jr./National Geographic Stock; 69 (UP), Victor R. Boswell, Jr./National Geographic Stock; 69 (LO), Victor R. Boswell, Jr./National Geographic Stock; 70, Richard T. Nowitz; 71, Stephen Uhraney, National Geographic My Shot; 72 (UP), Monica and Michael Sweet/Getty Images; 72 (LO), Joao Virissimo/Shutterstock; 73 (UP), Mike Theiss/National Geographic Stock; 73 (LO),

Fotosearch; 74, Anne Keiser/National Geographic Stock; 75 (UP), Jenni Allen/Shutterstock; 75 (LO), bikenbark/Bigstock. com; 76 (UP), Wikipedia; 76 (LO), Cindy Davenport/Shutterstock; 77 (UP), Wikipedia; 77 (LO), Norbert Rosing/National Geographic Stock; 78, Maria M. Mudd/National Geographic Stock; 79, Greg Dale/National Geographic Stock; 80, Brandon Cole; 82, Paul Souders/Getty Images; 83 (UP), Perry L. Aragon/ Getty Images; 83 (LO), Raul Touzon/National Geographic Stock; 84, Jarrud Knapp, National Geographic My Shot; 85 (UP), Bates Littlehales/National Geographic Stock; 85 (LO), Todd Gipstein/National Geographic Stock; 86, De Agostini/Getty Images; 87 (UP), Phil Weare/Getty Images; 87 (LO), Bianca Lavies/Getty Images; 88 (UP), Armin Rose/Shutterstock; 88 (LO), Christopher A. Klein/National Geographic Stock; 89 (UP), Gary Moss/Getty Images; 89 (LO), Richard A. Cooke/Corbis; 90 (UP), Paul Zahl/National Geographic Stock; 90 (LO), Victor R. Boswell, Jr./National Geographic Stock; 91 (UP), Paul Zahl/ National Geographic Stock; 91 (LO), Asahi Shimbun/Getty Images; 92, Darlyn A. Murawski/National Geographic Stock; 93 (UP), JonMilnes/Shutterstock; 93 (LO), David Liittschwager/ National Geographic Stock; 94 (LE), Inga Spence/Getty Images; 94 (RT), National Museum of Natural History, Smithsonian Institution; 95 (UP), Gary Meszaros/Getty Images; 95 (LO), Lynda Richardson/Corbis; 96, Steven Hayes/Getty Images; 97, Heather Perry/National Geographic Stock; 98 (UP), Joel Sartore/National Geographic Stock; 98 (CTR), William H. Crowder/ National Geographic Stock; 98 (LO), Fred Bavendam/Minden Pictures/National Geographic Stock; 99 (UP), Medford Taylor/ National Geographic Stock; 99 (LO), Jeff Foott/Getty Images; 100, Paul Nicklen/National Geographic Stock; 101 (UP), Raul Touzon/National Geographic Stock; 101 (LO), Joel Sartore/ National Geographic Stock; 102 (UP), James Forte/National Geographic Stock; 102-3, Joel Sartore/National Geographic Stock; 103, Dennis Kunkel Microscopy, Inc./Visuals Unlimited/ Corbis; 104, Frans Lanting//National Geographic Stock; 105 (UP), Raul Touzon/National Geographic Stock; 105 (LO), Norbert Wu/Minden Pictures/National Geographic Stock; 106, Flip Nicklin/Minden Pictures/National Geographic Stock; 107 (UP), Frank Russell/National Geographic My Shot; 107 (LO), Fred Bavendam/Minden Pictures//National Geographic Stock; 108, David Liittschwager/National Geographic Stock; 109, Daniel Gotshall/Getty Images; 110, Johann Schumacher/Getty Images; 112 (UP), Getty Images; 112 (LO), Bates Littlehales/National Geographic Stock; 113 (UP), Wikipedia; 113 (LO), Mark Beckwith/ iStockphoto.com; 114, George Grall/National Geographic Stock; 115 (UP), Kim Taylor/naturepl.com; 115 (CTR), Lynn M. Stone/ naturepl.com; 115 (LO), De Agostini/Getty Images; 116, Darrell Gulin/Getty Images; 117 (UP), Robert Shantz/Alamy; 117 (CTR), Johann Schumacher/Getty Images; 117 (LO), Don Farrall/Getty Images; 118, Carr Clifton/Minden Pictures/National Geographic Stock; 119, Carr Clifton/Minden Pictures/National Geographic Stock; 120 (UP), Visuals Unlimited, Inc./Alex Wild/Getty Images; 120 (LO), David A. Northcott/Corbis; 121 (UP), DEA Picture Library/Getty Images; 121 (LO), Scott Leslie/National Geographic Stock; 122 (UP), Visuals Unlimited, Inc./Alex Wild/Getty Images; 122 (LO), Visuals Unlimited/Corbis; 123 (UP), Charles Nesbit/Getty Images; 123 (LO), Joe McDonald/Corbis; 124 (LO), Will Heap/Getty Images; 124 (UP), Jay Greenberg; 125 (UP), Darlyne A. Murawski/National Geographic Stock; 125 (LO), Satoshi Kuribayashi/Nature Production/Minden Pictures; 126, Joel Sartore/National Geographic Stock; 127 (UP), Jef Meul/ Foto Natura/Minden Pictures/National Geographic Stock; 127 (LO), Hashime Murayama/Corbis; 128, Bjorn Van Lieshout/FN/ Minden Pictures/National Geographic Stock; 129 (UP), Mark Thiessen, NGP/National Geographic Stock; 129 (LO), Piotr Naskrecki/Getty Images; 130, Daniel Dempster Photography/ Alamy; 131, Tom Uhlman/Alamy; 132, Justin Lewis/Getty Images; 134, Greg Johnston/Getty Images; 135 (UP), Dorling Kindersley/ Getty Images; 135 (LO), Visuals Unlimited, Inc./Andy Murch/ Getty Images; 136, Monterey Bay Aquarium Foundation; 137 (UP), Jonathan Bird/Getty Images; 137 (LO), Hashime Murayama/National Geographic Stock; 138 (UP), Davies and Starr/Getty Images; 138 (LO), David Doubilet/National Geographic Stock; 139 (UP), Walter A. Weber/National Geographic Stock; 139 (LO), Gordon Wiltsie/National Geographic Stock; 140

(UP), The Washington Post/Getty Images; 140 (LO), Skip Brown/National Geographic Stock; 141 (UP), Lynda Richardson/Corbis; 141 (LO), Dorling Kindersley/Getty Images; 142, Kevin Fleming/Corbis; 143, Stephen Frink/Getty Images; 144 (UP), "adrisbow" (adriana lopetrone)/Getty Images; 144 (LO), John Cancalosi//National Geographic Stock; 145 (UP), Wikipedia; 146, George Grall/National Geographic Stock; 147 (UP), George Grall/National Geographic Stock; 147 (LO), Visuals Unlimited, Inc./Fabio Pupin/Getty Images; 148 (UP), Pete McBride/National Geographic Stock; 148 (LO), Tim Fitzharris/Minden Pictures/National Geographic Stock; 149 (UP), Walter A. Weber/National Geographic Stock; 149 (LO), MCT via Getty Images; 150, Tony Waltham/Robert Harding World Imagery/Corbis; 151, Allison Achauer/Getty Images; 152, Heidi and Hans-Jurgen Koch/Minden Pictures/National Geographic Stock; 154-155, Eric Isselée/Shutterstock; 154 (LO), Steve Winter/National Geographic Stock; 155, holbox/Shutterstock; 156 (UP), Brall Bralds/National Geographic Stock; 156 (LO), Mana Photo/Shutterstock; 157 (UP), Jim Richardson/National Geographic Stock; 157 (LO), Joel Sartore/National Geographic Stock; 158, Bill Curtsinger/National Geographic Stock; 159 (UP), Joseph T. Collins/Getty Images; 159 (LO), Joel Sartore/National Geographic Stock; 160 (UP), Brian Lasenby/Shutterstock; 160 (LO), David Wrobel/Visuals Unlimited, Inc./Getty Images; 161 (UP), Michael Durham/Minden Pictures/National Geographic Stock; 161 (LO), Michael Durham/Getty Images; 162, Tim Laman/National Geographic Stock; 163, Mike Theiss/National Geographic Stock; 164, Joel Sartore/National Geographic Stock; 165 (UP), Rich Phalin/iStockphoto.com; 165 (LO), Michael & Patricia Fogden/Minden Pictures; 166, Joel Sartore/National Geographic Stock; 167 (UP), Michael Nichols/National Geographic Stock; 167 (CTR), Universal Images Group Limited/Alamy; 167 (LO), Melinda Fawver/iStockphoto.com; 168 (UP), Joel Sartore/National Geographic Stock; 168 (CTR), Paul Sutherland/National Geographic Stock; 168 (LO), Don Johnston/Getty Images; 169 (UP), altrendo nature/Getty Images; 170 (UP), Michiel de Wit/Shutterstock; 170 (LO), George Grall/National Geographic Stock; 171 (UP), Cary Anderson/Getty Images; 171 (LO), Michael Durham/Getty Images; 172, Michael Durham/Minden Pictures/National Geographic Stock; 173 (UP), Raymond Gehman; 173 (LO), Karine Aigner/National Geographic Stock; 174, Greg Dale/National Geographic Stock; 175, Jeff Foott/Getty Images; 176, Klaus Nigge/National Geographic Stock; 178, Sebastian Kennerknecht/Minden Pictures/National Geographic Stock; 179 (UP), Robert Royse; 179 (LO LE), Joel Sartore/National Geographic Stock; 179 (LO RT), AvianArt Images by David Hemmings; 180, Glenn Bartley; 181 (LO), Glenn Bartley; 181 (UP), Andy Gehrig/iStockphoto.com; 182, Glenn Bartley; 183 (UP), Tim Laman/National Geographic Stock; 183 (LO), Jonathan Alderfer/National Geographic Stock; 184 (UP), Tim Fitzharris/Minden Pictures/National Geographic Stock; 184 (LO), Eric Isselée/Shutterstock; 185 (UP), Arthur Morris/Corbis; 185 (LO LE), Louis Agassi Fuertes; 185 (LO CTR), Allan Brooks/National Geographic Stock; 185 (LO RT), Allan Brooks/National Geographic Stock; 186 (UP), AvianArt Images by David Hemmings; 186 (LO), Robert Royse; 187 (UP), James Snyder, National Geographic My Shot; 187 (LO), Diane Pierce/National Geographic Stock; 188, Dickie Duckett/FLPA/Minden Pictures/National Geographic Stock; 189 (UP), FloridaStock/Shutterstock; 189 (LO), Robert Royse; 190, Chris Johns/National Geographic Stock; 191, Brechin Maclean/Getty Images; 192, Otto Plantema/FN/Minden Pictures/National Geographic Stock; 193 (UP), Michael O'Brien/National Geographic Stock; 193 (LO), John C. Pitcher/National Geographic Stock; 194 (UP), Blake Shaw/VIREO; 194 (LO), Robert Royse; 195 (UP), Thomas R. Schultz/National Geographic Stock; 195 (LO), Frans Lanting/National Geographic Stock; 196 (UP), Rudy Umans/Shutterstock; 196 (LO), Roy Toft/National Geographic Stock; 197 (UP), Paul Horsley/All Canada Photos/Corbis; 197 (CTR), Winfried Wisniewski/Getty Images; 197 (LO), Thomas R. Schultz/National Geographic Stock; 198, Gerry Ellis/Minden Pictures/National Geographic Stock; 199 (UP), Donald L. Malick/National Geographic Stock; 199 (LO), John & Barbara Gerlach/Getty Images; 200 (UP), George Grall/National Geographic Stock; 200 (LO), Visuals Unlimited, Inc./Robert Servranckx/Getty Images; 201 (UP), H. Douglas Pratt/National Geographic Stock; 201 (LO), Glenn Bartley; 202, Ingo Arndt/Getty Images; 203, John Raffaghello II/iStockphoto.com; 204, Image Source/Corbis; 206 (UP), Brian Skerry/National Geographic Stock; 206 (LO), Anne Keiser/National Geographic Stock; 207 (UP), Alaska Stock/National Geographic Stock; 207 (LO), Brian Skerry/National Geographic Stock; 208, Daniel McCulloch/Digital Vision; 209 (UP), Tory Kallman, National Geographic My Shot; 209 (LO), M. Watson/ARDEA; 210 (UP), Alaska Stock Images/National Geographic Stock; 210 (LO), Keenpress/National Geographic Stock; 211 (UP), Bates Littlehales/National Geographic Stock; 211 (LO), Hiroya Minakuchi/Minden Pictures/National Geographic Stock; 212, James Forte/National Geographic Stock; 213, Tim Fitzharris/Minden Pictures/National Geographic Stock; 214, Yva Momatiuk & John Eastcott/Minden Pictures/National Geographic Stock; 215 (UP), Louis Agassi Fuertes; 215 (LO), Ocean/Corbis; 216 (UP), Joel Sartore/National Geographic Stock; 216 (LO), Louis Agassi Fuertes; 217 (UP), Gerry Ellis/Minden Pictures/National Geographic Stock; 217 (LO), Joe McDonald/Corbis; 218 (UP), Karine Aigner/National Geographic Stock; 218 (LO), Tim Fitzharris/Minden Pictures/National Geographic Stock; 219 (UP), Phil Schermiester/National Geographic Stock; 219 (LO), Joel Sartore/National Geographic Stock; 220, Cary Anderson/Getty Images; 221, Luciana Whitaker/Getty Images; 222, Todd Tobey Photography/Getty Images; 224, Michael Melford/National Geographic Stock; 225 (UP), Bill Curtsinger/National Geographic Stock; 225 (CTR), Heidi and Hans-Jurgen Koch/Minden Pictures/National Geographic Stock; 225 (LO), Wikipedia; 226 (UP), Mauricio Handler/National Geographic Stock; 226 (LO), Andrew A. Martinez/Photo Researchers, Inc.; 227 (UP), Karen Kasmauski/Getty Images; 227 (LO), Carl & Ann Purcell/Corbis; 228, Richard Herrmann/Visuals Unlimited/Corbis; 229 (UP), Gary Braasch/Corbis; 229 (LO), Raul Touzon/National Geographic Stock; 230 (UP), Stephen St. John/National Geographic Stock; 230 (LO), Visuals Unlimited, Inc./Marc Epstein/Getty Images; 231 (UP), Tom Vezo/Minden Pictures/National Geographic Stock; 231 (LO), Steve Winter/National Geographic Stock; 232 (UP), Nik Wheeler/Corbis; 232 (LO), Arthur Morris/Getty Images; 233 (LE), MCT via Getty Images; 233 (RT), Nicole Duplaix/National Geographic Stock; 234, Raymond Gehman/National Geographic Stock; 235, Tony Arruza/Corbis; 236, Architect/Shutterstock; 237 (UP), Holmes Garden Photos/Alamy; 237 (LO), James P. Blair/National Geographic Stock; 238 (UP), Vincenzo Lombardo/Getty Images; 238 (LO), Mary E. Eaton; 239 (RT), Gerry Ellis/Minden Pictures/National Geographic Stock; 239 (LE), Melville B. Grosvenor/National Geographic Stock; 240, Medford Taylor/National Geographic Stock; 241 (UP), Thomas Mangelsen/Minden Pictures/National Geographic Stock; 241 (LO), Cary Sol Wolinsky; 242, Darlyne A. Murawski//National Geographic Stock; 243 (UP), David Sieren/Getty Images; 243 (LO), Helene Schmitz; 244, Tony Arruza/Getty Images; 245 (UP), George Steinmetz/National Geographic Stock; 245 (LO), Ocean/Corbis; 246, Denis Jr. Tangney/Getty Images; 247, Darlyne A. Murawski/National Geographic Stock; 253, John Coletti/JAI/Corbis; 254, Franco Cogoli/Grand Tour/Corbis; 255, Visuals Unlimited, Inc./Patrick Smith/Getty Images; 256 (LE), Photolibrary/Corbis; 256 (RT), Stacey Gold/National Geographic Stock; 257, Diane Cook & Len Jenshel/Corbis; 259, Mark A. Johnson/Corbis; 260, David Alan Harvey/National Geographic Stock; 263 (LE), jo Crebbin/Shutterstock; 263 (RT), Mike Brake/Shutterstock; 264, Frans Lanting/National Geographic Stock; 266 (LE), Bruce Dale/National Geographic Stock; 266 (RT), Alaska Stock Images/National Geographic Stock; 267 (LE), Tim Laman/National Geographic Stock; 267 (RT), Will Van Overbeek/National Geographic Stock; 268, William Manning/Corbis; 269, John Kernick/National Geographic Stock.

FIELD GUIDE TO THE
Water's Edge

STEPHEN LEATHERMAN AND JACK WILLIAMS

PUBLISHED BY THE NATIONAL GEOGRAPHIC SOCIETY
John M. Fahey, Jr., *Chairman of the Board and Chief Executive Officer*
Timothy T. Kelly, *President*
Declan Moore, *Executive Vice President; President, Publishing*
Melina Gerosa Bellows, *Executive Vice President; Chief Creative Officer, Books, Kids, and Family*

PREPARED BY THE BOOK DIVISION
Hector Sierra, *Senior Vice President and General Manager*
Anne Alexander, *Senior Vice President and Editorial Director*
Jonathan Halling, *Design Director, Books and Children's Publishing*
Marianne R. Koszorus, *Design Director, Books*
Susan Tyler Hitchcock, *Senior Editor*
R. Gary Colbert, *Production Director*
Jennifer A. Thornton, *Director of Managing Editorial*
Susan Blair, *Director of Photography*
Meredith C. Wilcox, *Administrative Director, Illustrations*

STAFF FOR THIS BOOK
Garrett Brown, *Editor*
Paul Hess, *Text Editor*
Sanaa Akkach, *Art Director*
Adrian Coakley, *Illustrations Editor*
Linda Makarov, *Designer*
Carl Mehler, *Director of Maps*
Michael McNey and XNR Productions, *Map Research and Production*
Jonathan Alderfer, *Contributing Writer*
Catherine Howell, *Contributing Writer*
Barbara Payne, *Contributing Writer*
Judith Klein, *Production Editor*
Mike Horenstein, *Production Manager*
Marshall Kiker, *Illustrations Specialist*
Britt Griswold, *Illustrator*

MANUFACTURING AND QUALITY MANAGEMENT
Christopher A. Liedel, *Chief Financial Officer*
Phillip L. Schlosser, *Senior Vice President*
Chris Brown, *Technical Director*
Nicole Elliott, *Manager*
Rachel Faulise, *Manager*
Robert L. Barr, *Manager*

The National Geographic Society is one of the world's largest nonprofit scientific and educational organizations. Founded in 1888 to "increase and diffuse geographic knowledge," the Society works to inspire people to care about the planet. National Geographic reflects the world through its magazines, television programs, films, music and radio, books, DVDs, maps, exhibitions, live events, school publishing programs, interactive media and merchandise. *National Geographic* magazine, the Society's official journal, published in English and 33 local-language editions, is read by more than 40 million people each month. The National Geographic Channel reaches 370 million households in 34 languages in 168 countries. National Geographic Digital Media receives more than 15 million visitors a month. National Geographic has funded more than 9,600 scientific research, conservation and exploration projects and supports an education program promoting geography literacy. For more information, visit www.nationalgeographic.com.

For more information, please call 1-800-NGS LINE (647-5463) or write to the following address:

National Geographic Society
1145 17th Street N.W.
Washington, D.C. 20036-4688 U.S.A.

For information about special discounts for bulk purchases, please contact National Geographic Books Special Sales: ngspecsales@ngs.org

For rights or permissions inquiries, please contact National Geographic Books Subsidiary Rights: ngbookrights@ngs.org

ISBN: 978-1-4262-0868-3
978-1-4262-0938-3 (regular)
978-1-4262-0939-0 (deluxe)

Printed in the United States of America

12/RRDW-LPH/1

SPECIAL SECTION

FOR BEACH LOVERS

All About Beaches

Beaches—the sandy playgrounds along our coasts—are America's number one recreational destination. Nothing restores the body and soul like a stay at the beach. We are naturally drawn to the rhythmic pounding of the waves as if we are returning to our primordial beginnings.

ANATOMY OF A BEACH

Every region—Atlantic, Pacific, or Gulf Coast—offers a different type of beach venue with distinctive recreational advantages. While swimming is the favorite activity of most beachgoers, many people just enjoy walking along beaches for the exercise and for the sounds, smells, and feelings that can only be sensed at the ocean's edge. Purists prefer wilderness beaches, but many vacationers are looking for creature comforts; the top city beaches offer all amenities. To choose the beach just right for you, there are a lot of factors to consider.

In scientific terms, beaches are accumulations of wave-deposited, loose sediment, usually sand, that extend from the outermost breaking waves to the landward limit of normal wave and swash action. Special

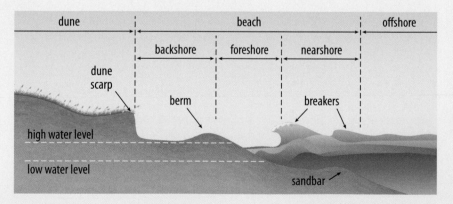

Lanikai Beach, a small beach on Oahu's North Shore, is known for green sea turtles.

words have been developed to name features in the profile of the beach as it slopes down from land to sea.

Wave action can cause dune scarping, or steep slopes. Berms are a major feature on most beaches; they separate the flatter backshore from the steeper seaward-sloping foreshore. The bar is an underwater ridge of sand, usually parallel to the shore, where incoming waves begin to break.

A large wave gives up its energy as it breaks on a Hawaiian beach.

Two youngsters enjoy a gentle surf as it comes ashore on a sandy beach.

WHAT'S A WAVE?

One of the special attractions of ocean beaches is the rhythmic, mesmerizing pounding of the waves. A person can sit on a beach for hours and enjoy the sheer power of waves as they give up their energy in spectacular displays upon breaking. Our concerns and problems seem to flow out with the tide as the wave action soothes and restores our spirits. Other people love to play in the waves and take advantage of their power. But we need to know how to recognize the dangers, such as shorebreak and rip currents.

Waves are undulating forms that move along the surface of the ocean. Primarily caused by wind blowing across the water, waves range in size from ripples caused by a gentle breeze to giant sea waves generated by hurricane-force winds. But all waves are similar, regardless of their size. They all have the features shown in the graphic below.

- **crest** high point of a wave
- **trough** low point of a wave
- **wave height** vertical distance from trough to crest
- **wave length** horizontal distance between adjacent crests
- **wave period** time in seconds for two adjacent wave crests to pass a fixed point

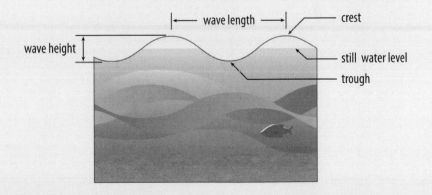

The surf crashes ashore at Cape Kiwanda—one of the best spots on the Oregon coast to experience spectacular wave action. Haystack Rock sits in the distance.

The 10,000 or so ocean waves that strike the water's edge every day are generated by the wind. As the force of the wind over the water increases, so does the surf. Waves breaking on a beach may have been generated locally, or they may have been spawned by storms at sea thousands of miles away.

Along North America's Atlantic coastline, hurricanes cause the largest waves. Along North America's Pacific coastline, migratory winter storms from high latitudes generate the largest swells.

Huge swells can crash onto beaches even in sunny, cloud-free weather because the storms that caused them are distant in both time and space. For example, giant ocean swells break on the north shore of Oahu, Hawaii, up to 40 feet high in January—hence the international surfing contests waged there in that season.

They that go down to the sea in ships,
that do business in great waters, these
see the works of the Lord, and his
wonders in the deep.
For he commandeth,
and raiseth the stormy wind,
which lifteth up the waves thereof.
—PSALM 107

WAVE MOVEMENT

Waves cause an oscillatory motion in the water, but no real forward movement of the water itself until they come close to shore and break. On a calm day with no breeze, a stick thrown in the water will move up and down as the waves pass, but it will not move horizontally on the water's surface. This phenomenon of wave motion is best illustrated by swaying seaweed caused by passing waves, which is indicative of the circular movement of the water particles.

When waves move into shallow water, the circular motion of the water particles cannot be maintained because of friction between the moving water and the sea bottom. Friction causes the lower part of the circular orbit of the wave to slow down relative to the upper part of the wave, exposed to the air. Finally, the orbit cannot be closed, and the top of the wave plunges forward and breaks. The forward movement of the water as it rushes up the beach face is called swash. This combination of physical processes results in what we call—and feel as—breakers, or waves breaking on the beach. Wave movement in shallow water also causes the sand to move, which is apparent from the turbid (sand-filled) water in the breaker zone.

Wave Speed

The speed of a wave is determined by the wavelength in deep water. The longer the wavelength, the faster the wave moves. Experienced sailors of yesteryear knew when a hurricane was nearby because of the arrival of the big swells.

In shallow water where the waves are feeling bottom (in other words, when friction between the wave motion and sea bottom is occurring), the water depth controls the speed of the wave. Therefore, waves slow down as they move into shallower water until they are forced to break. Waves steepen and grow higher as they start to break. Tsunami waves exhibit this characteristic in dramatic fashion as relatively low swells at sea become towering waves as they break on shore.

wave direction ⟶

Seaweed sways back and forth in response to the orbital movements of the water as a wave moves forward across the ocean.

Sea grass and kelp cling to a rocky coast.

A dolphin catches a ride on a breaking wave.

Birds soar above a breaking surf.

Clusters of bull kelp line a rocky shore.

TYPES OF WAVES

Breaking waves can be classified into three primary types: plunging, spilling, and surging.

∎ **Plunging waves** form when swells suddenly encounter a shallow bottom, such as a reef, a large sandbar, or a steeply sloping beach. The wave peaks and breaks suddenly, with all of its force concentrated in a limited area. Plunging breakers are the most spectacular as they fall forward and trap a cone of air that is compressed until it literally explodes. They often generate rip currents and shorebreaks on steep beaches and are responsible for many more injuries than spilling or surging waves.

This type of wave is most common along the Pacific coast, and it gives rise to the surfing tradition in southern California and Hawaii, especially the North Shore of Oahu. Large swells that encounter a reef or sandbar offshore become a surfer's delight as they break suddenly in a dramatic fashion.

A surfer catches a ride on a plunging wave, which are common on the Pacific coast.

A shorebreak is a wave that breaks directly onto a beach with great downward force. Swimmers—even experienced bodysurfers—need to beware of such waves, which are powerful and dangerous.

■ **Spilling breakers** are most common along the East and Gulf Coasts of the United States. They are formed when waves break over a long distance as the water becomes gradually shallower over a shallow sloping bottom. The breaking water rolls or tumbles forward, resulting in a wide and more gentle surf zone. Spilling waves generally are safe for wading, for children and inexperienced swimmers, and for novice bodyboarders. These waves are far less dangerous than plunging waves and can make for some good bodyboarding when the surf is really rolling ashore.

Spilling waves tumble ashore on an East Coast beach. These waves create a gentle surf.

Surging waves usually do not break but cause a sudden rise in the water.

■**Surging breakers** are less common than the other two types. They occur where the water is relatively deep near steep cliffs or coral reefs onshore, or at very steep rocky or pebbly beaches. Surging waves neither curl nor break, but they do cause a sudden rise and fall in water level. If not anticipated, these waves can cause serious injuries to people visiting rocky coastlines.

Men go forth to wonder at the heights of mountains, the huge waves of the sea, the broad flow of the rivers, the extent of the ocean, and the courses of the stars, and omit to wonder at themselves.

—ST. AUGUSTINE, *CONFESSIONS*

Waves can be dangerous, especially when they break onshore. Called shorebreaks, they occur when ocean swells hit steep beaches and abruptly pass from fairly deep water into very shallow water (see illustration, page 299). Shorebreaks rise quickly and break downward very hard. Daring swimmers may see shorebreaks as occasions for exciting bodysurfing, but they are the most dangerous waves at beaches, and even experienced surfers can be seriously injured when trying to ride them.

Waves often come in groups, which are called surf beat. A series of smaller waves builds up to the biggest breakers, and then the cycle repeats. Surfers wait for the biggest ones—usually the seventh to tenth waves on the southern California coast, for example—but when these big waves catch swimmers and waders unaware, they can be frightening. If you see one coming, the trick is not to jump over it but to take a deep breath and dive under it.

A sign warns visitors about the dangers of high surf.

BEWARE THE CURRENTS

Rip currents occur when large amounts of water are pushed far up the beach by the force of breaking waves, and then that water escapes back to the ocean in concentrated flows rather than broad sweeps. When a wave breaks, it produces swash—the word for the water that moves up and down the face of the beach. Bigger, taller waves propel the swash farther up the beach, and that propelled water naturally flows back down to the sea surface. Water follows the path of least resistance through holes or depressions in bars or along coastal engineering structures such as groins and jetties.

A strong, concentrated offshore flow is a rip current. (Riptide is a misnomer because tides have nothing to do with the generation of these powerful currents.) Rip currents typically flow seaward at two to three feet per second, but some megarips on the California coast have been clocked at six to eight feet per second. Most beachgoers will not (and should not) enter the ocean when waves are five feet or higher. Unfortunately many people drown on East and Gulf Coast beaches when the conditions seem safe: sunny, calm days with only two- to four-foot waves. Beachgoers need to learn how to read the waves and identify these dangerous and life-threatening currents.

Waves breaking at any angle to the shoreline result in longshore currents, which move you sideways, but not offshore into deep water. You may not even realize that there is a current until you look back at the beach for your towel, and then realize that it is tens to hundreds of feet down the beach. When large (especially plunging) waves break on the beach at an oblique angle, the longshore current can feel like a river flow.

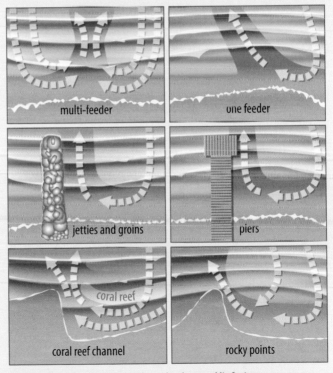

The shape of rip currents depends on the shore and its features.

FOR THE LOVE OF BEACHES

High waves can be dangerous. The energy they produce is proportional to the height of the wave squared. In other words, a three-foot wave is nine—not three—times more powerful than a one-foot wave. A general rule of thumb is that a surf with breaking waves five feet or taller is too dangerous for swimming. Such a situation may be good for surfing but not for swimming.

Learning & Loving the Waves

Waves are the heartbeat of the ocean, and one of the keys to pleasure and enjoyment at the ocean's edge. We are drawn to the rhythmic pounding of surf on beach or rocks and can spend hours gazing out and listening to the ever changing shore. The fresh, salty air is invigorating; the interplay of waves and shore captures the imagination and refreshes the psyche.

The better you understand the beach, its waves, and their potential, the more you can relish all that a beach has to offer. You can enjoy both the inner peace that comes with the quiet contemplation of gentle breakers and the wild exhilaration of big crashing waves. Waves and beaches are forever changing, erasing our footsteps in the sand, and cleansing our souls.

Tranquil waters and balmy breezes lure visitors to the beach to relax.

Palm trees on Oahu's Waimanalo Beach

River rocks at Maine's Acadia National Park

Sunset at Twin Rocks in Oregon

Lighthouse at Yaquina Head in Oregon

[What Makes a Beach Great?]

What does Dr. Beach consider when naming the Top 10 Beaches of the United States each year? Here are the 50 criteria he considers, followed by a chart you can use to rate your own favorite beach.

DR. BEACH'S RATING CRITERIA

■**Beach width at low tide** The wider the beach, the better. This means there is more beach to enjoy and allows for a wider range of activities, from volleyball and paddle tennis to jogging or beachcombing. If a beach is less than 10 meters wide, it ranks on the low end of the sliding scale. Anything as wide or wider than 100 meters scores a 5.

■**Beach material** Beach material is loose sediment deposited by waves. Fine sand is the most desirable beach material. Cobbles—naturally rounded stones larger than pebbles—rank at the other end of the sliding scale. Beaches with coarse sand score in the middle of the scale.

■**Beach condition or variation** Beach erosion is the wearing away of land and loss of sediments due to wave action and currents. Beaches that are prone to erosion slip in ranking. One that is depositional in nature—meaning sediment is deposited naturally—is more desirable and scores higher on the scale.

■**Sand softness** The softer and more powdery the sand, the more inviting the beach and the higher the rating. A light and fluffy sediment far outshines sand that is hard and dense. One advantage of a beach with more compact sand is that visitors can pack their bikes and enjoy a ride along the shoreline.

■**Water temperature** The most desirable beaches, particularly swimming beaches, have water in the range of 70 to

80°F. This makes for an inviting and refreshing swim in the surf. Senior citizens will probably want to add five degrees to that temperature, and kids are okay with temperatures five degrees cooler for short dips in the sea.

■ **Air temperature (midday)** At the beach, it is best for the air temperature to be warm enough for swimming and sunbathing. Thus, a midday temp of 80 to 90°F is best. A temperature that is lower than 60 or greater than 100 gets a low score.

■ **Number of sunny days per year** The more sunny days a beach can offer its visitors, the better. A beach in an area prone to cloudy and overcast skies means there is less time available to enjoy the sun.

■ **Amount of rain** To enjoy a beach and all it has to offer, a low amount of rain is best. In areas prone to high levels of precipitation, less time is available to enjoy the beach.

■ **Wind speeds** A gentle wind can offer some relief from the heat while you are at the beach, but high wind speeds are not ideal. They disrupt the day's experience by sending beach gear aloft and wind-blown sand into your eyes.

■ **Size of breaking waves** On a swimming beach, breaking waves can make bathing more exciting and interesting, but it is better for the waves to be on the low side and thus safer. Higher waves may be more dangerous and knock down a bather.

■ **Number of waves or width of breaker zone** The breaker zone is the zone within which waves start to break as they approach the coastline. A higher number of waves creates a more scenic and lively surf in which to swim and surf.

■ **Beach slope** Sand and wave size determine a beach's underwater slope. A gentle slope at the beach is much

A rainy day at the ocean

A starfish in crystal water

A tropical beach sunset

Debris cluttering a beach

A dog retrieving a toy

Maine's rocky coast

A kite aloft in offshore breeze

A scenic coastal vista

Fishing on a beach

A protected crescent-shaped beach   Snorkeling in crystal-clear water

preferred to one that is steep. It makes the water more easily accessible and safer, particularly for children.

■ **Longshore current** Along-the-shore currents are wave-generated currents that move parallel to the shore. A strong current can sweep swimmers into hazardous areas, such as piers, rocky zones, and coastal engineering structures like groins and jetties.

■ **Rip current** A rip current is a powerful channel of water that flows from the beach out into the ocean. It can be very dangerous for swimmers because it drags them away from the beach. Commonly occurring rip currents negatively impact a beach's score.

■ **Color of sand** The color of a beach's sand is determined by the source areas, such as coral reefs, rivers (past and present), volcanoes, or eroding sea cliffs. Whether pulverized lava, coral, or quartz, sand colors run the gamut from black to tan to pink to white. Pink or white sand beaches

Red tide caused by marine plankton

An open-air shower under a palm tree

score high. Beaches with gray sand score low. Light tan, the most common color, is in the middle of the pack.

■ **Tidal range** Large tidal ranges result in a lower score because of the potential danger of bathers being stranded on sandbars or shoals offshore as the tide rises.

■ **Beach shape** Beaches can range from pocket, or crescent shaped, to straight and elongated. Pocket-shaped sites score high because they provide more protection for swimmers. The water is apt to be calmer, hence safer in some places than along straight beaches

■ **Bathing area bottom conditions** The bottom conditions of bathing areas range from rocks and cobbles to mud and fine sand. Fine sand is most comfortable on the feet and best for navigating the ocean surf. A rock or cobblestone bottom makes walking more difficult, can be uncomfortable to the feet, and can even result in stubbing or cutting one's toe.

Snowy egret and its catch

A jellyfish awash on shore

■ **Turbidity** Ocean water that is murky, turbid, and lacking in clarity is not a welcome environment for beachgoers, especially those who want to bathe or swim. Water can be naturally turbid without being polluted. This occurs in areas with large expanses of marsh and large tidal ranges.

■ **Water color** Aqua blue water shining against a blue sky is most people's image of a perfect beach setting and is most inviting to swimmers. Tourmaline green water is gorgeous and also earns the highest mark; the color is caused by clear and relatively shallow water combined with highly reflective white sand underwater. A beach with grayish water scores at the bottom of the scale.

■ **Water quality** Clean water is the most important consideration for beachgoers because no one wants to recreate in a polluted area. Water quality is based on U.S. Environmental Protection Agency criteria.

■Red tide Red tide is seawater that is discolored by a large presence of marine plankton that produce toxins. It is best to frequent a beach where this never, or rarely, occurs, rather than one where it is commonplace.

■Smell Any kind of a rotten smell is going to send a beach's ranking to the bottom of the scale. Beachgoers seek fresh, salty air.

■Wildlife Plentiful fauna, from shorebirds to starfish, suggest a healthy shore environment and make for a more fulfilling experience for any beachcomber or nature lover. A lack of wildlife at the beach is often due to overdevelopment and loss of natural habitat.

■Pests Bothersome pests such as biting flies and mosquitoes can make a beach experience annoying and downright maddening. The presence of these pesky critters depends on prevailing winds, time of day and year, proximity to standing water, and nearby vegetation. A beach with few pests is most desirable.

■Sewage or runoff outfall lines The presence of any outfall lines at a beach severely hurts its desirability as a vacation destination. Such structures are unsightly, and they suggest hygiene issues that most people would want to avoid.

■Seaweed or jellyfish Swimming beaches that are free of bothersome seaweed and jellyfish, the latter of which can also sting (the Portuguese man-of-war is the worst of the lot on U.S. beaches), rank high. Common seaweed or jellyfish occurrences on the beach send the site to the bottom of the ranking.

■Trash and litter Beachgoers want a clean, pristine beach—a place where no trash can be found. Cigarette butts constitute the most common beach litter, while plastics make up the bulk of the material found on unkempt beaches.

■**Oil and tar balls** Tar balls are small, dark-colored clots of oil that can cling to your feet when you walk along the shore. Whether or not tar balls are present at a beach depends on several factors: tanker traffic, wind and sea patterns, and the incidence of oil spills.

■**Glass and rubble** The best beach scene is clean of any debris, particularly glass and rubble, or broken fragments of rock. Both not only make for difficult walking but also can cause injury.

■**Views and vistas (local scene)** An attractive, unob-structed local scene is a plus for a swimming beach and enhances any time spent on the coast. An obstructed view detracts from the scene and makes for a less desirable des-tination. High-rise hotels and condominiums can block the sun's rays during the afternoon along East Coast beaches in the United States.

■**Views and vistas (far vista)** Being able to look off in the distance and see an unhindered and stunning view—whether it be mountains, scenic islands with palm trees, or a beautiful horizon—enhances a day at the beach. A confined view scores low on the scale. Diamond Head seen from Hawaii's Waikiki Beach is the iconic far-beach vista.

■**Buildings and urbanization** A beach in a natural set-ting wins any day over a beach that is overdeveloped and urbanized with many buildings.

■**Access** Access to a beach is important for swimmers, fishers, and surfers who want to enjoy a day at the beach. Places that lack sufficient parking for the general public score low on the scale.

■**Misfits** No beachgoer enjoys a nearby misfit, such as a nuclear power station, an offshore dumping ground, or smokestacks. Any of these structures are clear black marks that send the beach score plummeting.

A kelp bed makes access difficult.

Wooden chairs invite beachgoers.

Smokestacks mar a beach scene.

Littered beaches deter visitors.

■**Amount of algae in water** Thick algae in beach water is unsightly and no fun to swim in, but it is only a nuisance. It is not harmful. Algae-free water scores high.

■**Nearby vegetation** Whether it is palm trees or dune grasses, a lot of vegetation not only creates a more visually appealing setting, but also signals a healthy ecosystem. In addition, trees can offer respite from the sun. Little or no vegetation as a result of overdevelopment sends a beach to the bottom of the scale.

■**Well-kept grounds and promenades or natural environment** A well-maintained beach makes for a much more pleasurable experience than one that is poorly tended. As such, it scores higher on the scale.

Lifeguards watch over a beach as clouds gather in the distance.

■**Amenities** A certain amount of amenities is a welcome presence on a beach. Bathrooms and showers are a must. Snack bars, shade, and chairs can enhance the beach experience.

■**Lifeguards** Lifeguards on a beach provide safety and security if an accident occurs or if a water rescue is necessary. When they are not present, you must swim at your own risk.

■**Safety record** Knowing the safety history of a beach is important before choosing it as a destination. If injuries or deaths have occurred, there is likely a safety issue that should be explored, and it drops to the lowest level on the scale.

■**Domestic animals** Not everyone is a dog lover—and not all owners pick up their pets' droppings—so the presence of dogs on the beach is a negative factor. A few domestic animals here and there are fine, as long as they are well tended. Some areas have designated dog beaches (apart from bathing and swimming beaches), where it can be fun to see the pets splashing in the water and enjoying themselves.

■**Ambient noise** No one wants to hear traffic noise from nearby highways or trains when spending a day at the beach. Peace and quiet are the criteria for a day of relaxing in the sun and communicating with nature.

■**Human noise** The prominent sound of a beach should be the surf, not boom boxes and rowdy parties.

■**Presence of seawalls, riprap, concrete, and rubble** Seawalls and riprap are used to stabilize shorelines and to help prevent erosional damage to roads and buildings. A beach with a lot of such coastal engineering structures slips down the sliding scale. Best are natural beaches that are not cluttered with concrete barriers.

■**Intensity of beach use** Beaches tend to attract lots of people, particularly during the height of the season. However, there is a tipping point beyond which a beach becomes too crowded. Much more welcoming is a site that offers ample open space in which to recreate.

■**Off-road vehicles** If off-road vehicles are allowed on the beach, it can mean a lot of noise, a danger to children, and general disruption of a truly relaxing experience. Ordinary autos allowed on a few beaches are just as disruptive. Daytona Beach, known for its beach driving, has banned driving in some areas.

■**Floatables in water** Floating litter in the surf relegates a beach to the bottom of the scale. A rare occasion is understandable, but no occurrence is best.

■**Public safety** Safety is always a concern, so it is worthwhile to find out if a destination beach has a history of crime. If so, it receives a low score. A rare incident is understandable.

■**Competition for free use** If the aim is a relaxing day at the beach—soaking up some sun and swimming in the surf—it is best not to have to compete with lots of fishers, boaters, or water-skiers and Jet Skiers for use of the beach. A little competition is acceptable as long as the water activities are well regulated and zoned by use for safety reasons.

A boardwalk provides access to a beach while protecting the area's vegetation and sand dunes.

RATE YOUR BEACH Using Dr. Beach's 50 Criteria

CRITERIA (1-25)	LEAST DESIRABLE	YOUR RATING	MOST DESIRABLE
BEACH WIDTH AT LOW TIDE	<10 m, narrow	① ② ③ ④ ⑤	>100 m, wide
BEACH MATERIAL	cobbles	① ② ③ ④ ⑤	fine sand
BEACH CONDITION OR VARIATION	erosional	① ② ③ ④ ⑤	depositional
SAND SOFTNESS	hard	① ② ③ ④ ⑤	soft
WATER TEMPERATURE	cold/hot	① ② ③ ④ ⑤	warm (70°–80°F)
AIR TEMPERATURE (MIDDAY)	<60° F, >100° F	① ② ③ ④ ⑤	80°–90°F
NUMBER OF SUNNY DAYS PER YEAR	few	① ② ③ ④ ⑤	many
AMOUNT OF RAIN	large	① ② ③ ④ ⑤	little
WIND SPEEDS	high	① ② ③ ④ ⑤	low
SIZE OF BREAKING WAVES	high/dangerous	① ② ③ ④ ⑤	low/safe
NUMBER OF WAVES/WIDTH OF BREAKER ZONE	none	① ② ③ ④ ⑤	6+
BEACH SLOPE (UNDERWATER)	steeply sloping bottom	① ② ③ ④ ⑤	gently sloping bottom
LONGSHORE CURRENT	strong	① ② ③ ④ ⑤	weak
RIP CURRENT	often	① ② ③ ④ ⑤	never
COLOR OF SAND	gray	① ② ③ ④ ⑤	white/pink
TIDAL RANGE	large (>4 m)	① ② ③ ④ ⑤	small (<1 m)
BEACH SHAPE	straight	① ② ③ ④ ⑤	pocket
BATHING AREA BOTTOM CONDITIONS	rocky, cobbles, mud	① ② ③ ④ ⑤	fine sand
TURBIDITY	turbid	① ② ③ ④ ⑤	clear
WATER COLOR	gray	① ② ③ ④ ⑤	aqua blue
WATER QUALITY (BEACH CLOSURES)	many	① ② ③ ④ ⑤	none
RED TIDE	common	① ② ③ ④ ⑤	none
SMELL (E.G., SEAWEED, ROTTING FISH)	bad odors	① ② ③ ④ ⑤	fresh salty air
WILDLIFE (E.G., SHOREBIRDS)	none	① ② ③ ④ ⑤	plentiful
PESTS (BITING FLIES, MOSQUITOES)	common	① ② ③ ④ ⑤	no problem

The higher the total, the better the beach.

CRITERIA (26-50)	LEAST DESIRABLE	YOUR RATING	MOST DESIRABLE
PRESENCE OF SEWAGE/RUNOFF OUTFALL LINES ACROSS THE BEACH	several	① ② ③ ④ ⑤	none
SEAWEED/JELLYFISH ON THE BEACH	many	① ② ③ ④ ⑤	none
TRASH AND LITTER	common	① ② ③ ④ ⑤	rare
OIL AND TAR BALLS	common	① ② ③ ④ ⑤	none
GLASS AND RUBBLE	common	① ② ③ ④ ⑤	rare
VIEWS AND VISTAS – LOCAL SCENE	obstructed	① ② ③ ④ ⑤	unobstructed
VIEWS AND VISTAS – FAR VISTA	confined	① ② ③ ④ ⑤	unconfined
BUILDINGS/URBANISM	overdeveloped	① ② ③ ④ ⑤	pristine/wild
ACCESS	limited	① ② ③ ④ ⑤	good
MISFITS (NUCLEAR POWER STATION, OFFSHORE DUMPING)	present	① ② ③ ④ ⑤	none
ALGAE IN WATER AMOUNT	infested	① ② ③ ④ ⑤	absent
VEGETATION (NEARBY). TREES, SAND DUNES	none	① ② ③ ④ ⑤	many
WELL-KEPT GROUNDS/PROMENADES OR NATURAL ENVIRONMENT	no	① ② ③ ④ ⑤	yes
AMENITIES (SHOWERS, CHAIRS, ETC.)	none	① ② ③ ④ ⑤	some
LIFEGUARDS	none	① ② ③ ④ ⑤	present
SAFETY RECORD (DEATHS)	some	① ② ③ ④ ⑤	none
DOMESTIC ANIMALS (E.G., DOGS)	many	① ② ③ ④ ⑤	none
NOISE (CARS, NEARBY HIGHWAYS)	much	① ② ③ ④ ⑤	little
NOISE (E.G., CROWDS, RADIOS)	much	① ② ③ ④ ⑤	little
PRESENCE OF SEAWALLS, RIPRAP, CONCRETE/RUBBLE	large amount	① ② ③ ④ ⑤	none
INTENSITY OF BEACH USE	overcrowded	① ② ③ ④ ⑤	ample open space
OFF-ROAD VEHICLES	common	① ② ③ ④ ⑤	none
FLOATABLES IN WATER (GARBAGE)	common	① ② ③ ④ ⑤	none
PUBLIC SAFETY (E.G., CRIME)	common	① ② ③ ④ ⑤	rare
COMPETITION FOR FREE USE OF BEACH	many	① ② ③ ④ ⑤	few

Beach Buzz

92 TERMS TO DESCRIBE BEACHES

■**accretion** Deposition of sediment, usually sand, which is evident by the seaward advance of a shoreline indicator, such as the high water line, berm crest, or vegetation line. Accretion causes the beach to widen. Opposite of erosion.

■**aeolian** Referring to the transport and deposition of sand by wind, which is the principal means by which sand dunes are formed.

■**armoring** Placement of fixed engineering structures, typically rock, wood timbers, or concrete, on or along the shoreline to reduce coastal erosion. Armoring structures include seawalls, revetments, bulkheads, and riprap.

■**backshore** Generally dry portion of the beach between the berm crest and the vegetation line that is submerged only during high water levels and eroded during storms.

■**backwash** The seaward return flow of swash on the beach face due to gravity.

■**bar** Submerged mound of sand that generally runs parallel to the shore and causes waves to break before reaching the beach.

■**barrier beach** A low-lying, sandy island or spit that lies offshore and is generally parallel to the mainland.

■**beach** Accumulation of wave-deposited, loose sediment, usually sand, that extends from the outermost breakers to the landward limit of wave and swash action.

■**beach loss** Volumetric loss of sand, usually measured by a loss of dry beach width.

■**beach monitoring** Periodic collection of data, such as dry beach width, to study changes over time.

■ beach narrowing Decrease in usable (dry) beach width caused by episodic storm impact or long-term erosion.

■ beach nourishment Sand artificially placed on the beach, usually by pumping sea-bottom sediments onshore, to replace the sand being lost alongshore or offshore. Beach nourishment projects are usually large in scale, spanning many miles of shoreline to rebuild eroded beaches.

■ beach profile Measurement of the elevation or height of the beach surface taken along a line that runs from the dune to the water across the beach. Profiles taken at different dates can be compared to illustrate and quantify storm, seasonal, and longer-term changes in beach width, height, volume, and shape.

■ berm Feature usually located at mid-beach and characterized by a sharp break in slope, separating the flatter backshore from the seaward-sloping foreshore.

A surfer rides the curl of a mammoth breaking wave.

∎**blowout** Small, often circular or oval depression in sand dunes, caused by wind scouring where protective vegetation has been disturbed.

∎**bluff** High, steep bank or cliff of noncoastal origin along the mainland. Steepened bluffs are caused by wave undercutting of the cliff toe.

∎**breakwater** Structure built parallel to the shoreline and seaward of the beach, designed to protect the beach and upland areas by causing waves to break and dissipate their energy before reaching the shore. Breakwaters are constructed of large boulders or concrete tetrapods.

The sea, after all, is a living "minestrone," home to small, medium, and sometimes very large creatures, each one a microcosm of chemicals acting on the surrounding sea. In short, the ocean is not only filled with life; its very nature is shaped by life.
—SYLVIA A. EARLE AND LINDA K. GLOVER,
OCEAN: AN ILLUSTRATED ATLAS

∎**building setback** State or locally required seaward limit of beachfront construction, usually for a house.

∎**bulkheads** Rigid structures with vertical walls built parallel to the shoreline to serve as barriers to wave attack and to prevent storm surge flooding of upland areas; constructed of treated wood, corrugated steel, PVC, or other materials.

∎**coastal compartment** Stretch of shore that is connected by a common longshore sediment transport system, such as the south shore of Long Island, New York.

■**cusps** Crenulated beach surfaces characterized by an evenly spaced series of rounded, small headlands (projections) and bays (or embayments). The alongshore spacing of cusps ranges from a few feet to hundreds of feet, and their relief varies from a few inches to several feet.

■**deflation** Lowering of beach profile usually caused by sand blowing away.

■**dune** Mound or ridge of sand that is deposited by the wind and capable of movement when unvegetated. Dune building can be augmented by sand fencing or planting beach grass.

■**dune restoration** Technique of rebuilding an eroded or degraded dune through one or more methods (sand fill, fencing, revegetation, etc.).

■**dune walkover** Light construction (usually elevated) that provides pedestrian access across a dune without trampling the vegetation.

■**ebb current** Tidal current moving away from the coast during a falling (ebbing) tide, often with high-velocity flows through tidal inlets.

■**ebb tidal delta** Sandy shoals formed by ebbing currents found on the seaward side of tidal inlets.

■**erosion** Physical removal of beach sand that is transported offshore, alongshore, or into bays and lagoons via inlets. Erosion results in shoreline recession—landward retreat of a shoreline indicator such as the high water line, vegetation line, or dune line. Opposite of accretion.

■**erosion hot spots** Areas where erosion is occurring at a much higher rate than those of adjacent beach areas. This can threaten beachfront development or infrastructure, especially during storms.

■**erosion watch spots** Areas where the coastal environment (natural or built) will soon be threatened if shore erosion trends continue.

■**eustatic sea-level rise** Worldwide changes of sea level over decades to centuries caused by addition of water from the melting of glacial ice and/or thermal expansion of seawater due to global warming.

■**fetch** Distance of open water over which the wind blows in the development of waves. The fetch length can restrict wave development so that only relatively small waves occur in narrow bays and lagoons.

■**flood current** Tidal current moving toward the shore through a tidal inlet or up a tidal river, estuary, or lagoon.

■**flood tidal delta** Sandy shoals formed on a rising (flooding) tide and found on the estuarine or lagoon side of a tidal inlet.

■**foreshore** Seaward sloping portion of the beach within the normal range of tides.

■**geotextile tubes** Elongated cloth bags or tubes made out of plastic material that can be stacked or arranged as a form of semihard coastal engineering.

■**groins** Shore protection structures that extend from the beach backshore into the surf zone, perpendicular to the shoreline. A groin is intended to build up an eroded beach by trapping littoral drift or to retard the erosion of a stretch of beach. Often misidentified as jetties.

■**hard stabilization** Emplacement of treated wood, rocks, concrete, PVC, and/or steel in the form of breakwaters, bulkheads, groins, jetties, seawalls, etc.

■**high water line** The line separating wet from dry sand and formed by swash uprush on the beach face.

A high bluff on the coast

Fishing coastal waters

A tranquil beach scene

A starfish on a beach

Kayaking river rapids

Coastal salt marshes

A large tree claimed by erosion

A family playing on a beach

A river flowing into the ocean

A wooden dock on a lakeshore at sunset

Playing in breaking waves

■**hurricanes** Tropical cyclones that involve winds 75 miles per hour or faster, spiral inward toward a core of low pressure, and rotate in a counterclockwise direction in the Northern Hemisphere.

■**isostatic** Describing local or regional changes in the ground surface elevation, resulting in land subsidence or uplift.

■**jetties** Shore-perpendicular structures built at the sides of an inlet to maintain navigable waterways. They stabilize an inlet by intercepting the longshore transport of sand that would otherwise fill it in or cause the channel to shift position. Jetties are often confused with groins, but they are much longer and more substantial structures and are usually built in pairs.

■**littoral budget** Sediment budget of the beach, consisting of sources and sinks.

■**littoral drift** Sand and coarser material moved in the breaker and swash zones by waves and longshore currents along the shoreline.

Black lava cliffs on the Hawaiian coast

A shore-protection structure on a beach

■**littoral system** Area from the landward edge of the coastal upland (usually the dune) to the seaward edge of the nearshore zone.

■**longshore current** Current moving parallel to the shore and generated by waves breaking at an angle to the shoreline.

■**longshore sediment transport** Sediment transport along the beach (parallel to the shoreline) caused by longshore currents and/or waves approaching obliquely.

■**mean sea level** The average elevation of the sea surface, determined from tide gauges.

■**neap tide** Small tide range, occurring at the first and third quarters of the moon, when the gravitational pull of the sun opposes that of the moon.

■**nearshore** Underwater area close to the beach and often characterized by sandbars, where sediment is actively being moved by waves and currents. This zone

A woman and her dog stroll along a beach at low tide.

typically extends to a depth of 25 to 30 feet along the Atlantic coast.

■**nodal point** Location of longshore sediment transport divergence, where the littoral drift moves away in opposite directions along the coast. Normally areas of higher erosion rates.

■**nor'easters** Extratropical storms with winds that commonly blow from the northeast, occur during the winter, and can generate large waves and elevated tides, resulting in considerable beach and dune erosion.

■**oblique wave approach** Waves that approach the beach at an angle (i.e., not straight on) and generate longshore currents.

■**offshore** Seaward of the nearshore zone, where sediment transport is initiated only by large swells or coastal storms.

■**overwash** Wave uprush overtopping the beach and dunes during storms; water and entrained sand that are moved landward of the dune. Also called an overwash surge during major events.

■**peat** Dark brown to black fibrous material produced by plants that grow in marshes or bogs. When exposed on the beach face, peat indicates long-term erosion and landward barrier migration.

■**perigean** Describing the twice-yearly period of time when the moon is at its closest approach to Earth, and the tidal range is larger than normal.

■**perigean spring tides** Coincidence of perigean and spring tidal conditions resulting in the highest high and the lowest low tides. Nor'easters, such as the Ash Wednesday Storm of 1962, become even more damaging when they occur during perigean spring high tides.

■**recession** Landward movement of the shoreline due to the loss of beach material and/or direct inundation of the land.

■**refraction** The bending of waves by bars and shoals that can concentrate wave energy on a portion of the shoreline, resulting in accelerated beach erosion.

■**relative sea level rise** The gradual rise in the water level relative to the land surface due to worldwide changes in the volume of seawater and/or local vertical movement of the land.

■**revetment** Facing of stone, concrete, or rubble built to protect an embankment or upland against erosion by wave action or currents.

■**ridge** A longshore feature that may become exposed at low tide; often formed by a bar moving onshore as a form of post-storm beach recovery.

■**rip current** Strong, localized current flowing seaward from the shore; sometimes visible as an agitated band of water, which is the return movement of water piled up on the shore by incoming waves. Rip currents are by far the biggest killers of ocean swimmers.

■**riprap** A layer or protective mound of stones placed randomly to prevent erosion of upland areas. Also the name of the stone so used.

■**runup** Part of swash action caused by breaking waves.

■**sand bags** Sand-filled cloth or geotextile bags that can be stacked to provide semihard coastal protection and are designed to retain sand while allowing water to flow through them.

■**sand waves** Much larger features than cusps that may migrate along the shoreline. Sand waves can locally cause accelerated erosion, known as erosion hot spots. Also called shoreline meanders, sand humps, or giant beach cusps.

■**scarp** Vertical drop-off of the dry beach caused by oblique wave attack during stormy conditions; beach scarps can be several inches to over six feet high and can disappear by the return of sand onshore during berm accretion. Dunes can also be scarped, forming vertical, wave-cut faces.

■**scarping** Erosion of a dune or berm, usually by oblique wave attack during a storm.

■**scour** Removal of beach sand or rocks by waves and currents such as at the base of a dune or the toe of a shore structure.

■**seawalls** Vertical or near-vertical shore-parallel structures designed to prevent upland erosion and storm surge flooding. Seawalls are generally massive concrete structures emplaced along a considerable stretch of shoreline at urban beaches.

■**shadow effect** Stretch of sand-starved, eroded beach that is downdrift of a structure such as a jetty or groin and hence in the littoral drift "shadow" of that structure.

■**shoal** A large deposit of sand, generally created by currents near inlets, that can be an obstruction to boats and can cause wave refraction.

■**shoreline** Boundary between the land and the sea, which is often defined as the mean high water line for mapping purposes.

A couple walks along a seawall in British Columbia. Seawalls help prevent erosion of upland areas.

■**soft stabilization** Artificial emplacement of sand via beach nourishment or by building and enhancement of sand dunes with sand fencing or vegetative plantings. Sand scraping of the beach to build up sand dunes is another means of soft stabilization.

■**sort** Separate particles into various-size categories by moving water or wind.

■**spoil** Dredged sediment, usually from inlets or lagoons, that can be clean or polluted.

■**spring tide** Larger than average tidal range that occurs twice monthly during new and full moons.

■**storm surge** Sudden, temporary rise of sea level primarily due to winds but also caused by atmospheric pressure reduction, resulting in piled-up water against the coast, which is the primary cause of coastal flooding during a storm.

■**swash** Sheet of water that flows up and down the beach foreshore due to breaking waves and gravity, respectively.

■**swell** Long-period waves that tend to widen the dry beach, usually in summer months or during fair weather.

■**tidal inlet** Channel through a barrier beach, which is characterized by swift currents that interrupt the littoral drift of sand.

■**tidal range** Difference in height between high and low tide.

■**undertow** General layman's term used to describe nearshore currents that may "suck" swimmers underwater. A more accurate description is backwash from large breaking waves, not to be confused with rip currents.

■**updrift** Opposite to the direction of the littoral drift's predominant movement. Opposite of downdrift.

■**uprush** The movement of water (swash) up the beach face when a wave breaks on the foreshore.

■**washover or washover fan** Sand deposited during storms landward of the dune line, sometimes extending to the marshes or into the bay waters.

■**wave height** Vertical difference between a wave's crest and trough; higher waves are more energetic and can cause rapid beach changes.

■**wavelength** Distance between successive wave crests.

■**wave period** Time in seconds between successive wave crests. Swells are long period, while sea waves are short-period.

■**wave refraction** Bending of waves along their length because of shallow bars and sand shoals.

An updated birding field guide is something to chirp about!

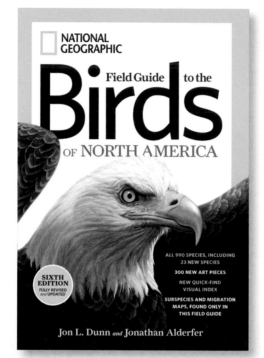

2012 © National Geographic Society